Genetic Ancestry

Genetic Ancestry focuses on the scientific nature and limitations of genetic ancestry testing. Co-authored by a genetic anthropologist and a cultural anthropologist, it examines the social, historical, and cultural dimensions of how people interpret genetic ancestry data. Utilizing examples from popular culture around the world and case studies from the Caribbean, the chapters highlight how genetic technology can sometimes bolster racial thinking while also serving as a tool of resistance and social justice.

Jada Benn Torres is a genetic anthropologist, Associate Professor and Director of the Genetic Anthropology and Biocultural Studies Laboratory in the Department of Anthropology at Vanderbilt University, USA.

Gabriel A. Torres Colón is a cultural anthropologist, Assistant Professor in the Department of Anthropology and Assistant Director of the American Studies Program at Vanderbilt University, USA.

New Biological Anthropology

Series Editor: Agustín Fuentes, Princeton University, USA

Emergent Warfare in Our Evolutionary Past
Nam C. Kim and Marc Kissel

The Promise of Contemporary Primatology
Erin P. Riley

Genetic Ancestry
Our Stories, Our Pasts
Jada Benn Torres and Gabriel A. Torres Colón

For more information about this series, please visit: www.routledge.
com/anthropology/series/NBA

Genetic Ancestry
Our Stories, Our Pasts

**Jada Benn Torres and
Gabriel A. Torres Colón**

Routledge
Taylor & Francis Group

LONDON AND NEW YORK

First published 2021
by Routledge
2 Park Square, Milton Park, Abingdon, Oxon OX14 4RN

and by Routledge
52 Vanderbilt Avenue, New York, NY 10017

Routledge is an imprint of the Taylor & Francis Group, an informa business

British Library Cataloguing-in-Publication Data
A catalogue record for this book is available from the British Library

Library of Congress Cataloging-in-Publication Data
Names: Torres, Jada Benn, author. | Torres Colón,
Gabriel A., author.
Title: Genetic ancestry : our stories, our pasts / Jada Benn
Torres and Gabriel A. Torres Colón.
Description: Abingdon, Oxon ; New York,
NY : Routledge, 2021. |
Series: New biological anthropology | Includes
bibliographical references and index.
Identifiers: LCCN 2020020696 (print) | LCCN 2020020697
(ebook) | ISBN 9780367026240 (hardback) | ISBN
9780429398650 (ebook)
Subjects: LCSH: Physical anthropology. | Genetic genealogy.
Classification: LCC GN60 .T67 2021 (print) | LCC GN60
(ebook) | DDC 599.9—dc23
LC record available at https://lccn.loc.gov/2020020696
LC ebook record available at https://lccn.loc.gov/2020020697

ISBN: 978-0-367-02624-0 (hbk)
ISBN: 978-0-429-39865-0 (ebk)

Typeset in Times New Roman
by codeMantra

We dedicate this book to those who have dared to dream about themselves in new and more powerful ways and to those on a quest to reimagine the past and work towards a better future.

Contents

Series editor foreword

The human experience continually presents circumstances wherein the biological and the social, the historical and the possible, the knowable and the yet-unimagined, are enmeshed, entangled, and dynamically interlaced. Genetic ancestry is certainly one of those instances.

Genetic ancestry as a concept is multimodal; as a methodology, it is often at the cutting edge of technoscience, and as a practice, it is intricately intertwined with race, racial experience, and racism. How one thinks about, defines, and uses it makes a difference in the social fabrics they weave. The academy, and the world beyond it, needs a thoughtful, accurate, and anthropological overview of how genetic ancestry is created, assessed, analyzed, and manipulated. This is that book.

In these pages, Jada Benn Torres and Gabriel A. Torres Colón demonstrate that genetic ancestry is a biocultural phenomenon that involves techno-methodology, scientific inquiry, and social creativity, and that varies cross-culturally. Setting the stage by immersion in culture, identity, and ancestry, and the details of analytic of the DNA testing itself, they confront the "elephant in the room," race and racism, head on. Their attention focuses on the often hyper-racialized societies of the Americas, where genetic ancestry connects explicitly to how people think about and experience race. Benn Torres and Torres Colón invite us to reconceptualize race as "embodied difference," forcing the reader to dive deep into the what, where, when, and why of human relatedness. They offer us a view of genetic ancestry that is "more than a tool for making biological connections between people, living and dead, near and far." They don't de-biologize it; rather, they illustrate how it is a technology, and a biopolitic, that can be used to acknowledge and make sense of "the stories that people create about themselves, their place in the world, and about others."

One reads this book and is wholly convinced that ancestry is a fundamentally social concept, and the science and technologies of genetic ancestry are only truly empirically and anthropologically sound when contextualized in the social, cultural, and experienced worlds and lives of those whose DNA is being sequenced. Engaging kinship, relatedness, and the questions of ancestry offers simultaneously simple and hyper-complex scenarios, and a myriad of "correct" answers. Benn Torres and Torres Colón interweave diverse strands of the biological and sociocultural anthropological literatures, showing why (and how) the answers to the questions "who are your ancestors?" are "complex and never straightforward with the overlapping and shifting notions of the stories that people tell about themselves and their pasts."

This is not to say there is no biologically relevant information in genetic ancestry testing. There is a "there" there in the material (DNA) being measured. The results do produce biological information (data). However, ancestry testing is a methodological tool, not a magic bullet or crystal ball into the truth of who individuals are and where they come from. The data emerging from such tests must be analyzed, and that analysis can be framed, used, and structured in multiple ways. Genetic ancestry tests can ascertain information about ancestry based on biological notions of relatedness, but these are only one of many ways in which people relate to one another. And such information never exists in a vacuum. Comparing individuals' sequences of DNA to reference populations tells us little by itself, but placed in dialogue with cultural, historical, and lived sources of information, it speaks volumes. To clarify why and how this is so, Benn Torres and Torres Colón integrate kinship; nationalism; embodied difference; and a myriad of notions of identity, self, and community with a deep discussion about the science and the modes of analyses involved in genetic ancestry. In doing so, they reveal that there is no one "natural" relationship between genetic ancestry data and how people interpret genetic information, but there are multiple genuinely experienced modes of such engagement.

Drawing on their own, and anthropologist Sarah Abel's, work, they immerse the reader in the lived experiences of what they are framing, injecting ethnographic richness into the narrative. North American African Americans, Afro-Brazilians and the Brazilian mestiçagem, French Antilleans, Puertorriqueños, and Maroons in rural Jamaica breathe life into the theoretical and the conceptual. These examples show how individuals, communities, and societies give, receive, and co-construct meaning with genetic ancestry results and practices via the dynamic lived processes of ancestry, race, and racial experience. They illustrate how genetic ancestry can be used as a form of empowerment in marginalized and racialized communities so much more than a biomarker of lineage bio-kin. They offer an expansion to the standard tropes of genetic ancestry and its relationships to socially constructed race, and the structural violence of racism by encouraging the reader to take on board the concept of "embodiment of difference" as biocultural processes related to racism/racial experience. In doing so, they offer insight into the dynamic range of relationships between genetic ancestry and racial experience as an alternative to the tired and often racist trope of genetic ancestry as racial identifier or constructor.

This book is an important contribution to anthropology and beyond in the technical and conceptual explorations and insights Benn Torres and Torres Colón provide, but it also does another form of critical work, one that many talk about, but few actually do: decolonizing the canon. Throughout the text, authors from inside and outside the traditional canon(s) in genetic anthropology and biological and social-cultural anthropology co-exist. By including voices from an array of sources and placing them in dialogue with the largely white (and often male) perspectives typical of much anthropology and genetics theory, Benn Torres and Torres Colón force the reader to see that powerful ideas and important theoretical frames are found across multiple intellectual perspectives and experiences. A primary reliance on the core "standards" often silences the very voices and perspectives that we need, now, in these discourses.

For example, Benn Torres and Torres Colón offer a whole chapter inviting the reader to engage with the Haitian scholar Anténor Firmin and his deployment of a biological anthropology as anti-racism scholarship. In 1885, Firmin published *The Equality of the Human Races*, a monumental critique of Gobineau's earlier "Essay on the Inequality of Human Races" and of the broader racist anthropology of the time (and of now). Benn Torres and Torres Colón justifiably assert "if anthropologists claim Boas as the father of American anthropology for the four-field approach that is exemplary exhibited in 'Mind of Primitive Man', then anthropologists might have a long-lost grandfather in Haiti." They show why acknowledging his presence and entering into a dialogue with our ancestor Firmin, and many others, can decolonize the canon by expanding it, offering the possibility of a better, more "positivist" (as per Firmin's book's subtitle) anthropology.

In these pages, Benn Torres and Torres Colón elegantly and emphatically demonstrate that genetic ancestry is increasingly part of our stories and our past ... sometimes an important part and sometimes not important at all. This narrative lays clear why an integrative anthropology is the necessary home for scientific inquiry into biocultural dimensions of race and racial experience, and that such a perspective acts to augment and enhance an anti-racist anthropology. As one navigates this text, they are struck, with intellectual clarity, by the truth that "genetic ancestry has no inherent social meaning, and that we must pay careful attention to how people think about the past and relatedness before we can understand how people experience genetic ancestry technology and information."

This book is indeed an example of a New Biological Anthropology, replete with rich ethnographic, genomic, historical, and sociocultural entanglements. This is an anthropology that is inclusive of robust biological as well as social sciences and one that is fundamentally anti-racist.

Agustín Fuentes
Cleveland, OH, USA
April 2020

Acknowledgements

There are many people whose support and encouragement were crucial in the development of this book. None of this work would be possible without our community partners and interlocutors in Puerto Rico, Jamaica, St. Vincent, and Trinidad. Their interest, time, and engagement with our work has been so formative in our thinking about ancestry that without their intellectual and experiential inputs our work would not have been possible. We are thankful to every participant that entrusted us with their DNA and their knowledge about their histories and communities. We also acknowledge those who did not want to participate. Their refusals often involved conversations that helped us to more fully recognize and respect the guiding principle of autonomy and to keep that principle central in our research.

Agustín Fuentes provided us with the opportunity, encouragement, and support to write a book that was theoretically integrative of both our diverse anthropological backgrounds and research experiences. Of course, his own work in advancing a more generous, integrative, and biocultural anthropology has been foundational in our own thinking.

Our good friend, Thomas Kies, who has been providing meaningful feedback since our graduate school days at the University of New Mexico, took on the difficult task of providing feedback with his sharp scholarly eye and outstanding pedagogical perspective. If this book is accessible to a broad audience, Tom's feedback was instrumental in that effort.

To the College of Arts and Science, the Office of Equity, Diversity and Inclusion, and our colleagues at Vanderbilt University, we are grateful for the infrastructural and financial support that was necessary to see our research projects and this book to fruition.

Our colleagues in the Department of Anthropology and American Studies at Vanderbilt have been supportive of our fieldwork, engaging with our ideas, and accepting of our collaboration. Too often, academia is not entirely welcoming of co-authored works; but we know this bias is misguided. In fact, this work has been every bit as challenging as our single-authored works. It is easy to make calls for integrative anthropology and a return to four-field anthropology—actually doing these things requires difficult intellectual work that is only possible with support from colleagues and institutions.

Finally, our parents—Elba, Luis, and Janice—have given us the love (and childcare!) necessary to have the privilege of blending our family with our fieldwork. To Jada's father, Clyde, though it has been 20 years since he died, the memory of his love and curiosity about our ancestors provided the intellectual seeds for this work. Our kids, Savion and Lureida, who have been awesome fieldwork companions since they were toddlers, are the reason we get up to work and find purpose in our research.

Introduction

A controversial DNA test

In response to President Donald Trump's challenge to prove her Native American ancestry, Elizabeth Warren—Senator from Massachusetts and 2020 presidential candidate—produced a DNA test as proof of a family history claiming Cherokee and Delaware ancestry. It is difficult to believe that Sen. Warren anticipated the backlash from various Native American tribes, and specifically citizens of the Cherokee Nation. In an editorial for *Tulsa World*, Secretary of State of the Cherokee Nation Chuck Hoskin Jr. explained why he and others had rebuked Sen. Warren's effort.

Mr. Hoskin begins with the analogy of friends and family gathering around the table for a meal. This analogy is used to explain the concept of citizenship to non-Native friends. Family and friends are welcomed:

> However, if something tragic were to happen to me or my wife, not everyone at that table would be entitled to inherit my house or become legally responsible for my kids. Those friends at my table will no doubt be critical in healing and providing love for those left behind, but it is my immediate family who will have their own rights and responsibilities in the eyes of the law.
>
> (Hoskin 2019)

If families are like citizens, Mr. Hoskin argues, then friends are political allies who need to understand that "they are fighting for our citizens' benefits, not their own." Mr. Hoskin has no problem with family stories of Native ancestry, but he points to the complex contemporary and historical reality of people in the United States who have tried to take advantage of claimed Native ancestry for financial benefit. Furthermore, tribal citizenship laws have their origins in historical times when Native peoples continued to experience centuries of marginalization. Therefore, Mr. Hoskin concludes:

> When someone boasts they are Native American due to the results of a DNA test, it perpetuates the general public's misunderstanding about what it means to be a tribal citizen … When the national dialogue is focused on inaccurate concepts of citizenship and hurtful name-calling, we ask that it be focused on what citizenship truly means and how we can all make Native America stronger.
>
> (Hoskin 2019)

Native American scholars also weighed in the controversy. In a public statement, Dr. Kim TallBear characterized Sen. Warren's test as "yet another strike—even if unintended—against tribal sovereignty" (Schilling 2018). She points to the power inequality that genomic scientists hold in the United States and how Sen. Warren's reliance on a DNA silenced Native American objections to outside attempts to define their identity and citizenship laws. In an editorial published in the *Atlantic*, Diné (Navajo) geneticist Krystal Tsosie (2018) also argues against the relationship between DNA and Native identity and added scientific nuance to the interpretation of Warren's test as linking her to indigenous ancestry but not to a specific tribe. In another piece that Tsosie coauthored with Matthew Anderson (2018), they point to the history of Native American distrust of genetic researchers, which has led to less participation of Native Americans in research and a detachment from genetic technology to determine tribe membership. All of these claims by Native political leaders and scholars are made in light of a long history of Native Americans being colonized, socially marginalized, and politically silenced.

Yet, despite these clearly articulated positions by Native Americans, popular news outlets published opinion pieces seeking to provide other perspectives to the controversy. Science journalist for *The New York Times*, Carl Zimmer, weighed in by clarifying the science behind DNA tests in order to conclude the following:

> What these results will mean to Senator Warren's sense of herself is for her to work out. For her part, she has said that she would not claim membership in any tribe based on these results. Everyone else getting ancestry tests also will have to decide what their results mean for themselves. But if we want to come to a full reckoning with our DNA, we have to get a lot more familiar with the science behind it.
>
> (Zimmer 2019)

NBC News fact-checked the controversy, and like many other news outlets, distinguished between Sen. Warren's claim of ancestry and her own claims to not being a tribal citizen, having Native identity, or thinking of herself as a racial minority. With that background, NBC News then asks, "So, is Warren actually Native American?" Before contextualizing the scientific limits of the ancestry test, they answer, "Her DNA strongly suggests it" (Timm 2018). Taking yet another political angle, Jennifer Bendery of the HuffPost published a news report and opinion piece titled, "Mainstream Media Is Blowing Its Coverage Of Elizabeth Warren's DNA Test: Tribal Leaders and Native People Say the Senator Is an Ally—and They Support Her Look at Her Ancestry. But Hardly Anyone Asked Them" (2019). Bendery reached out to elected tribal leaders, to whom she presumably gives importance as elected representatives of many Native Americans, and argued that many Native Americans supported Warren; furthermore, the few voices cited in pieces by *The New York Times* were not representative of all Native Americans.

Dr. Kim TallBear was one of the cited Native Americans in Bendery's piece. Bendery characterized TallBear as a known Native critic of Elizabeth Warren. TallBear dismissed Bendery's piece as one that used Native people as instruments in a political discussion that was both exclusive of Native Americans and meant to be consumed by non-Natives. The Native American Journalist Association (NAJA) asserted that the HuffPost and Bendery have "oversimplified a complex topic that is critically important to Indigenous communities" (najastrategy 2019). NAJA identifies the central question in the controversy: "Is Warren's claim and the roll out of the DNA test results harmful to Native people and their political and cultural identity?" They point out that Bendery's coverage missed important details when dismissing Cherokee Nation Secretary of State Chuck Hoskin Jr. as a nonelected official and diminishing his voice as marginal relative to other Native leaders who supported Warren. Most importantly,

> The idea that a handful of Indigenous people can speak for the majority is deeply rooted in hurtful stereotypes, colonial attitudes and ideas of racial superiority. Indigenous communities often hold conflicting viewpoints on important issues and show concern about multiple matters affecting their lives.
>
> (najastrategy 2019)

On February 1, 2019, multiple news outlets reported that Sen. Warren had apologized to the Cherokee Nation.

Our stories, our past

This book aims to provide a broad conceptual framework for thinking about genetic ancestry. We embrace anthropological and critical race theory perspectives in examining how humans think about their past through and beyond genetic ancestry. The central proposition in this book is that in order to understand how humans experience genetic ancestry, we must understand (1) the science behind genetic ancestry, (2) the multiple ways in which humans think about their ancestors, (3) the social, cultural, and intellectual contexts in which humans experience genetic ancestry, and (4) the ways in which race shapes how we think about ancestry and social differences. These are a lot of factors to consider, but we think each is intricately related to one another.

The controversy over Sen. Elizabeth Warren's DNA test involves science, politics, cultural differences, and history. At first, we might be tempted to grapple with the scientific validity of genetic ancestry science as Carl Zimmer of the *New York Times* did. However, even after we figure out if the *methods* employed by Dr. Carlos Bustamante in conducting Warren's DNA test are scientifically valid (and we think they are), we have to consider that scientists can disagree on the interpretation of genetic data. Krystal Tsosie specifically criticizes Bustamante's report for using the term "Native American" instead of "indigenous" when referring to Warren's ancestor. Tsosie's criticism is twofold. First, indigenous peoples from throughout the Americas refer to themselves in many ways (e.g., "First People," "Indian," "Native Americans"), so Bustamante's use of "Native American" obfuscates the diverse ways that indigenous peoples identify themselves. Second, scientists make estimates about an individual's ancestry by comparing their DNA to a referent population—in this case, indigenous populations from somewhere in the Americas. Tsosie takes Bustamante to task for the ways in which diverse indigenous groups are lumped together and utilized as references for indigenous populations throughout the United States. Given Tsosie's criticisms, it becomes apparent that we need additional historical and political contexts, in order to fully appreciate the meaning and impact of genetic ancestry testing.

Warren's family stories about a Native ancestor are usually the sort of historical evidence that we can use to make sense out of a genetic result. Some Native American scholars and political leaders, like Tsosie and Hoskin Jr., claim that without genealogical proof, Warren's indigenous ancestor could come from anywhere in the Americas. But pinpointing where from the Americas an indigenous ancestor comes from is too high a of historical burden of proof for meaningfully interpreting genetic ancestry. Warren's family history and known genealogy makes it more difficult to explain how her indigenous ancestor could have come from elsewhere in the Americas than coming from North America. It is impossible to know with absolute certainty, but scientific and historical truths are rarely "known" with absolute certainty. Nevertheless, insights about scientific certainty are not the end of this controversy. Scholars like Tsosie and TallBear are well aware of how genetic ancestry tests work—their objections are not just scientific, they are also cultural and political. In fact, both Tsosie and Tall-Bear are essentially educating their audience about the inseparability of science, culture, and politics.

Claiming ancestors is a political act for many Native American tribes. By "political," we mean that there are struggles over citizenship, autonomy, and social policy that are tied to questions of who gets to claim Native American ancestry. These are political questions that Native Americans have struggled with over many years of colonization and exclusion from U.S. politics. In turn, these processes are tied to how race and racism have been a central form of social organization in the United States and the Americas. Therefore, claiming any form of ancestry is a political act because it is political for Native Americans. Native American politics are historically and contemporaneously tied to national U.S. politics; so, when a national politician like Warren makes a claim to Native American ancestry, it is inevitably tied to Native American politics. The initial claim by Warren and her supporters that Warren was not claiming Native ancestry or tribal citizenship is beside the point because, as all the Native people cited above unequivocally state, there is no way of making such claims to Native ancestry without political consequences. The only way we can assert that Warren's claim is not politically relevant to Native Americans is by excluding Native American political concerns from our public conversations.

The notion that this controversy should not really concern Native Americans because Warren was simply responding to Trump is extremely problematic because it essentially suggests that there can be major political debates concerning Native American ancestry that excludes Native Americans from the conversation. To be sure, there have been Native American leaders talking about these issues for a long time and their voices are intellectually and politically diverse. However, their existence has not translated to a significant presence in the U.S. public sphere. Tsosie and TallBear are two scholars who happen to have expertise on genetic science and Tsosie is a geneticist. There are more, but there are certainly not enough Native American scholars to significantly influence how we ask and answer questions about ancestry in the U.S. public sphere. If we want to better understand genetic ancestry, then we need to question the sources of knowledge and how the terms of the debate include and do not include people from marginalized communities. For the world of research exists in a broader social world, so issues of race and equity in how we carry out our scholarship are very much related to our actual scholarship. To put it differently, the quality of science and scientific interpretation is tied to the social dynamics of scientific practitioners.

How this book is organized

Our Stories, Our Past presents a conceptual framework for dealing with the social, cultural, and scientific complexities and intertwining of genetic ancestry in today's world. Our main argument is that genetic ancestry is a biocultural phenomenon that varies cross-culturally. In the context of the Americas, which is our main area of study, genetic ancestry as a biocultural phenomenon is contiguous with how people think about and experience race. We propose to think about race as something that we call "embodied difference," which immediately forces us to think about race in relation to gender and other forms of human relatedness. Moreover, we are adamant that thinking about genetic ancestry as biocultural and through embodied difference also means that we must pay attention to broad frameworks of scientific practice and social politics. These frameworks are not the musings of scholars from racialized communities thinking about science in relation to race; instead, science and politics, as we will show, can be prisms through which we can improve scientific practice. We are advocating better science, not the subversion of science through reflection and politics. Once scientific practice is sharpened through the lenses of scientists from racialized communities, we can imagine ways in which genetic technologies can be utilized by racialized marginalized communities as a form of resistance to racial oppression—exactly the opposite of what happened in the Warren controversy.

The book is divided into six chapters. The first three chapters serve as an introduction to the fundamental concepts and arguments that we embrace in the study of genetic ancestry. Chapter 1 explains the basic science of genetic ancestry tests and research, while Chapter 2 explores various social and cultural dimensions of how humans think about ancestry. These chapters are geared toward providing a brief introduction to important concepts for readers that might be less familiar with biological anthropology or cultural anthropology. In Chapter 1 we present some of the key concepts underlying genetic ancestry testing including population genetic and statistical principles of ancestry testing. In Chapter 2, we examine how time and space affect various cultural understanding of the past, which is an important dimension for understanding ancestry and ancestors. Furthermore, we examine some of the most fundamental analytical prisms for thinking about relatedness. Our discussion of kinship and nationalism cover fundamental anthropological approaches; however, we introduce the concept of "embodied difference" as a way to think about relatedness and racial experience. This discussion about embodied difference, in turn, lays important conceptual groundwork for examining the relationship between genetic ancestry and race in subsequent chapters. Chapter 3 surveys the literature and our own research about how genetic ancestry exists in various social settings. Our message is rather straightforward: genetic ancestry does not have the same cultural impact in every society. Accordingly, we argue that the social impact of genetic ancestry should be an empirical question that examines genetic technology relative to sociocultural context.

The last three chapters examine how race, as an articulation of embodied difference, is a social and cultural phenomenon that consistently and significantly impacts how people experience difference, think about their ancestors, and formulate scientific anthropological inquiry, especially throughout the Americas. Chapter 4 offers a biocultural and pragmatic theory of race that demonstrates how genetic ancestry can be both biologically and culturally meaningful. While we fully support anthropological efforts to undermine biological arguments for racial difference, we maintain that how people experience race complicates social scientific claims that race is just a social construct. Chapter 5 dives into an intellectual historical examination of Anténor Firmin, a nineteenth-century Haitian intellectual who utilized scientific anthropology in anti-racist and anti-colonial struggles. When compared with other important anti-racist anthropological intellectuals in history, Firmin's legacy can help us appreciate the indispensability of science as an instrument for anti-racist scholarship. We make this argument by demonstrating how "anti-racist science" is, in fact, nothing less than a more empirically and theoretically robust form of science. Chapter 6 brings together the lessons from the previous chapters and draws on our research in Jamaica and Puerto Rico to demonstrate how genetic ancestry can and has served as a tool for empowerment by racially marginalized communities.

References

Bendery, Jennifer. 2019. "Mainstream Media Is Blowing Its Coverage of Elizabeth Warren's DNA Test." *Huffpost*, January 4, 2019. https://huffpost.com.

Hoskin Jr., Chuck. 2019. "Chuck Hoskin Jr.: Elizabeth Warren Can Be a Friend, but She Isn't a Cherokee Citizen." *Tulsa World*, January 31, 2019. https://tulsaworld.com.

najastrategy. 2019. "NAJA Calls Huffington Post Reporting Irresponsible—Native American Journalists Association." January 9, 2019. https://najanewsroom.com/2019/01/09/naja-calls-huffington-post-reporting-irresponsible/.

Schilling, Vincent. 2018. "Strike Against Sovereignty? Sen. Warren Asserts Native American Ancestry via DNA." *Indian Country Today*, October 15, 2018. https://indiancountrytoday.com.

Timm, Jane C. 2018. "Fact Check: Elizabeth Warren Took a DNA Test. Does It Prove She's Native American?" *NBC News*, October 16, 2018. https://www.nbcnews.com/politics/donald-trump/fact-check-trump-wants-warren-prove-her-native-american-heritage-n864446.

Tsosie, Krystal. 2018. "Elizabeth Warren's DNA Is Not Her Identity." *The Atlantic*, October 17, 2018. https://theatlantic.com.

Tsosie, Krystal and Matthew Anderson. 2018. "Two Native American Geneticists Interpret Elizabeth Warren's DNA Test." *The Conversation*, October 22, 2018. https://theconversation.com.

Zimmer, Carl. 2019. "Opinion | Before Arguing about DNA Tests, Learn the Science Behind Them." *The New York Times*, October 18, 2019. https://www.nytimes.com/2018/10/18/opinion/sunday/dna-elizabeth-warren.html.

1 What is a genetic ancestry test?

Within the last 20 years, direct-to-consumer (DTC) tests, including genetic ancestry tests, have become more accessible to a wide market of consumers. With DTC testing, the consumer purchases the services of a company that screens the consumer's DNA for particular markers related to a trait of interest or, in the case of ancestry testing, assesses the consumer's DNA for presumed ancestral origins. Unlike conventional genetic testing that may be ordered for medical or legal reasons, DTC testing is a transaction between the testing company and the consumer, without direct medical or other third-party interventions (Norrgard 2008). There are several different types of DTC genetic tests available, including paternity tests; ancestry tests; and tests for athletic ability, mate choice, or even wine preference (Phillips 2016). These types of DTC tests are marketed as non-medical products and consequently can be understood as a form of "recreational" genetics in which the consumer is free to act, or not, on the information.

In the late 1990s the first DTC company came onto the commercial market and this fundamentally changed the way in which consumers began to interact with their DNA. In 1997, the first DTC genetic tests were sold by a now non-existent company called GeneTree (Rosen 2003). This company initially focused on primarily paternity tests then later expanded to other types of genetic tests including ancestry tests. In 2001, GeneTree was sold to Sorenson Molecular Genealogy Foundation, based in Salt Lake City, and GeneTree's services were absorbed into the emerging genetic genealogy business. Within a decade of the founding of GeneTree, the DTC genetic testing market was booming (Borry, Cornel, and Howard 2010). Currently, growth in the DTC genetic testing sector appears to be moving upward with recent market value estimates of around $140 million but projected to hit $340 million by 2022 (Statista 2017). Though not the most popular testing option (that distinction goes to paternity tests), genetic ancestry testing is becoming more accessible and more common among consumers. According to a February 2019 *MIT Technology Review* article, over 26 million people worldwide have taken a genetic ancestry test and in 2018, more people purchased genetic ancestry tests in that year than all previous years combined (Regalado 2018, 2019).

Given the amount of attention and money that genetic testing in general has garnered, a number of researchers have begun to study the DTC testing phenomenon and are finding some interesting trends. Their studies suggest that the appeal of DTC tests can be generalized into three main categories: (1) identity-seeking, (2) disease risk, and (3) lifestyle/curiosity (Pearson and Liu-Thompkins 2012; Saukko 2013; Su 2013). The popular press is also cued into the growing interest in genetic testing evident from the now commonplace headlines warning about the limitations of genetic tests, privacy issues surrounding testing, as well as (and discussed in the Introduction) the social ramifications of genetic results. This growth in the number of consumers seeking to explore their own genomes and the consequential growth of the DTC market indicate an increased public engagement with genetic technologies. However, despite the interest in genetics and its potential to shed light upon a test-taker's identity, disease risk, or lifestyle, several researchers have found that many potential DTC genetic test consumers lack a fundamental understanding of genetics, how these tests work, and the interpretations of test results (Lachance et al. 2010; Leighton, Valverde, and Bernhardt 2012; Pearson and Liu-Thompkins 2012). In response to this disconnect, in the remaining parts of this chapter, we lay out the mechanics behind one type of DTC test, genetic ancestry testing, with the intent to provide a brief but thorough overview of how and why genetic ancestry tests operate.

Basic science behind genetic ancestry tests

A genetic ancestry test is a quantitative estimation of a test-taker's genetic background, which links a test-taker to populations that are associated with broad geographic regions. Here, genetic background refers to the compilation of genetic variation that has been shaped by both biological and environmental factors. For the purposes of ancestry testing, genetic backgrounds are tied to populations that reside in specific geographic regions and shared genetic background is assumed to be indicative of shared ancestral origins. In other words, with genetic ancestry tests, test-taker's patterns of genetic variation are compared to groups of people of known ancestral origin. We call these groups of people of known origin reference populations. When there are matches between the test-taker and the reference populations, the test-taker's ancestry is assumed to be among the matching reference populations. These comparisons between test-takers and reference populations are based on a series of well-established analytical methods, which will be described later in this chapter. Since the quality of any ancestry test is dependent on the comparisons between a test-taker and reference population, reference populations must be carefully defined. If reference populations are inappropriately defined or if inappropriate reference populations are used in ancestry tests, the results will not be informative about the test-taker's actual ancestry.

In general, reference populations are usually drawn from populations that are believed to have been located in the same area of the world for extended periods of time, that is, have not moved out of their traditional residence area, and have experienced long periods of relative genetic isolation, meaning they have not incorporated other populations into their population. In reality, no population strictly meets these requirements, as people have moved across the globe and have continuously mixed with other peoples they have encountered. However, drawing on the ways in which human genetic variation is distributed across geographic space geneticists can use sophisticated statistical models to make probabilistic assessments of relatedness between a test-taker and reference population. These results should then be interpreted with proper contextualization such as documented or oral histories, in order to be most informative about a test-taker's ancestral origins.

Just as it is important to identify appropriate reference populations for ancestry tests, it is also of paramount importance to use appropriate parts of the genome to discern ancestry. In genetic ancestry tests, a series of genetic markers known as ancestry informative markers (AIMs) are used to distinguish the relationship between a test-taker and reference populations. A genetic marker describes a place in the genome that you are interested in and you know its location within the genome. Genetic markers can have alternate forms known as variants. Genetic variants can be found on every chromosome. For example, on chromosome one, there is a genetic marker called rs2814778. This genetic marker is within the DARC (FY) gene and this gene is involved in producing a protein that can be found on the surface of red blood cells. While everybody has this genetic marker in their genome, there are different forms or variants of this marker and the variants can differ between individuals. AIMs are genetic markers whose variants are unequally distributed across human groups, where some variants are common in some populations and the same variant is rare in other populations. A genetic marker that is recognized as an AIM can arise randomly within a population; this process is known as genetic drift.

Alternatively, AIMs can also come about due to natural selection. In the case of natural selection, the variant may be a genetic adaptation to something in the local environment. The genetic marker mentioned above, rs2814778, is actually an AIM. There are two variants of this genetic marker: one variant produces the red blood cell protein while the other variant is the null variant, meaning it does not produce the protein. When this protein is on the surface of a red blood cell, it can be used by a malaria parasite, *Plasmodium vivax,* to enter the red blood cell. Once the parasite enters the red blood cell, it causes the person to be infected with malaria. Individuals that have the null variant of rs2814778 do not make this protein and as a result, the malaria parasite cannot enter the person's blood cell and the person does not become ill. Those with the null rs2814778 variant are generally immune to infection with *P. vivax*, though they can still get malaria from a different malarial parasite (Hodgson et al. 2014). The null variant is therefore considered a genetic adaptation in response to the *P. vivax* parasite (Kwiatkowski 2005). This particular genetic adaptation is actually one of several genetic adaptations to malaria that have been observed across human groups (Luzzatto 2012; Fan et al. 2016; Mackinnon et al. 2016). The rs2814778 null variant is in fact very common in individuals from regions of the world where malaria is endemic, but specifically in West and Central Africa where virtually all individuals have this particular genetic adaptation. Outside of Africa where this type of malarial parasite has not been as problematic, virtually nobody has the rs281477 null variant as a genetic adaptation. Because of the particular distribution of the rs2814778 variants, this genetic marker can be used as a clue to an individual's ancestry and is therefore considered an AIM. As part of an ancestry test, when the rs2814778 null variant shows up in a test-taker, it is indicative of ancestry from West or Central Africa (Kano et al. 2018). Contemporary genetic ancestry tests utilize upward of 500,000 markers, inclusive of AIMs and other variants, from across the entire genome (Jobling, Rasteiro, and Wetton 2016). Through the use of these large panels of genomic markers, ancestry tests then rely on statistical algorithms to assess proportional ancestry.

Conceptual framework: the distribution and structure of human genetic variation

From a conceptual standpoint, ancestry tests are based upon a characteristic distribution of human genetic variation across geographic space. This characteristic distribution is clinal in nature, meaning that there are no absolute boundaries in the distribution of different genetic variants. Instead, genetic variants are spread across geographic areas in a continuum where variants of one kind are more common in some areas of the world and then gradually become less common in other parts of the world. This clinal distribution of genetic variants happened because over the course of human history, people have tended to find mates that are geographically close to them. Consequently, populations that are geographically near to each other tend to exchange mates and as a result tend to share variants. On a global scale, this has resulted in a direct relationship between genetic and geographic distances between populations. In other words, the further away populations are from one another the less genetically related they are while conversely, populations that are geographically close to one another are genetically close as well.

Despite the clinal nature of the distribution of human genetic variation, researchers can assess variation in a particular way so that human variation appears to be ordered by homogenous sub-groupings of populations. This type of ordering in which there are sub-groupings within a population is referred to as population substructure. Population substructure is most notable in human groups when geographically distant groups are considered. The genetic differences seen between groups that are geographically distant are the result of different environments, inclusive of different selective pressures, that shape genetic variation. Recall, the example of the rs2814778 variants where the presence of an environmental stressor, the *P. vivax* malarial parasite, resulted in a genetic adaptation seen primarily only in individuals with ancestry from regions that are affected with endemic malaria. In environments where there is no endemic malaria this genetic adaptation is not commonly observed in people and consequently using the rs2814778 variants becomes useful for distinguishing between ancestry from sub-Saharan Africa and regions outside of sub-Saharan Africa. This relationship between environment, genetics, and geography contributes to population substructure. Genetic ancestry tests rely on population substructure, or detecting homogeneous sub-groupings, usually from geographically distant populations to estimate the ancestry of test-takers. In genetic ancestry results, this corresponds to the percentage of ancestry from broad continental regions, for example, sub-Saharan Africa, Europe, Asia, among others.

Within the last few years, however, researchers have made a variety of methodological and analytical changes to the kind and number of markers used in ancestry tests in order to assess finer levels of substructure within geographic areas. In the broadest terms, these methodological and analytical changes refer to improved sampling of reference populations, incorporating additional genetic markers, as well as more sophisticated statistical models for assessing ancestry. For example, Leslie et al. (2015) examined population substructure across the UK for over 2,000 people that had all four grandparents from the UK as well as an additional 6,000 people from across continental Europe. Using over 500,000 genomic markers, they could discern several ancestral clusters indicative of major events within the history of Britain including the initial peopling of the British Isles, Roman rule, as well as the influx of Norse and Danish Vikings (Leslie et al. 2015). This high-resolution analysis allowed Leslie and colleagues to examine genetic perspectives of the ancestry and demographic histories of the UK. Studies like Leslie et al. (2015) illustrate the power of large-scale, high-resolution genetic analyses to learn more about the pre- and more recent history of human groups. Similar types of methods are being used in commercial ancestry tests in order to discern finer-scale genetic ancestry estimates for individual test-takers.

Assessing genetic ancestry: local and global ancestry

As mentioned in a previous section of this chapter, genetic ancestry tests are probabilistic assessments that illuminate the relationship between a test-taker and reference populations by relying on population substructure observed within and between populations. Genetic ancestry tests can assess ancestry from two main vantage points: local ancestry and global ancestry. Local ancestry assesses the ancestry of segments of an individual's chromosomes and is useful for understanding which parts of the genome were inherited from particular ancestral groups. Accordingly, local ancestry accounts for the mosaic nature of a test-taker's ancestry where an individual has inherited sections of their genome from ancestors that came from different regions. However, not every test-taker will have diverse ancestry, in which case, local ancestry will indicate that much of each chromosome was inherited from a singular or a limited number of regions in the world. Arguably, for the average ancestry test-taker, local ancestry becomes a useful tool for visualizing how ancestors contributed to their current genetic makeup. As will be discussed at the end of Chapter 3, the recognition of a mosaic ancestry underscores the disassociation between the race and genetic ancestry.

Global ancestry estimates are the general proportional origins of an individual test-taker. This estimate assesses an individual's general ancestry by averaging ancestral proportions across the genome so that the total proportion adds to 100% (Y. Liu et al. 2013). This estimate is interpreted as the percentage of the test-taker's genome that was inherited from ancestors associated with a particular geographic region (Thornton and Bermejo 2014). For the purposes of exploring family history using genetic genealogies, most test-takers are primarily interested in global ancestry estimates because these estimates provide the most comprehensive results about ancestry from broad geographic regions and consequently are potentially useful for examining genealogies and family oral histories.

Though both local and global genetic ancestry results can provide some general assessment about a test-taker's ancestral origins, there are some important caveats in making sense of ancestry results. For example, some test-takers test with multiple companies in hopes of corroborating the ancestry results across testing companies. However, due to differences in reference populations used by each testing company as well as differences in the algorithms used to make ancestry estimates, each company can provide different ancestry estimates. The different estimates from each testing company should not be considered as proof of the unreliability of ancestry testing in general but, taken in whole, can be used to gain a general appreciation of the most likely regions of ancestry for the test-taker. Likewise, test-takers may also be interested in validating their ancestry estimates by comparing their ancestry estimates to the ancestry estimates of a sibling. This type of comparison is also likely to result in different estimates. Every person inherits roughly half of their DNA from each parent. However, siblings do not necessarily inherit the same half of DNA from a parent as their sibling. This results in each sibling inheriting a different complement of genomic markers, and this complement of genomic markers shows a different relationship to reference populations resulting in different ancestry estimates. While differences in ancestry estimates should not be an issue for identical twins, in practice, identical twins can get slightly different ancestry estimates (Agro and Denne 2019). These differences in estimates can be due to genotyping anomalies, which can then affect the statistical analysis that underlies the ancestry estimates. For example, if there were differences in which genetic markers were successfully genotyped, then identical twins could end up with different panels of genetic markers that are fed into the ancestry estimation algorithm. These algorithms can then come up with different ancestry estimates despite the identical nature of twins' DNA. Overall, these sorts of issues are not indications that genetic ancestry tests are inherently imprecise but is in fact a testament that ancestry estimates are just that—estimates. This point leads us to the next topic on genetic ancestry testing—statistical analysis.

Statistical aspects of genetic ancestry estimation

Ancestry estimation is a statistically and computationally complex process. At a minimum, around 40 AIMs with strong discerning power are considered sufficient to identify broad continental origins (Hoggart et al. 2003; Rosenberg et al. 2003). Additional AIMs and other genomic markers are added to genetic ancestry estimations in order to improve the quality and resolution of the ancestry estimates (Price et al. 2010; J. Liu et al. 2013). Some of the most popular ancestry testing companies use several dozen reference groups and upward of a half-million genomic markers in their tests to identify more specific continental regions of origin for their test-takers. Consequently, these types of tests require high-performance computing in order to make the ancestry estimates (Jobling, Rasteiro, and Wetton 2016). There are several analytical approaches used to produce ancestry estimates. These approaches differ in their underlying models that describe how reference groups contribute to descendant groups. In addition, there are differences between how the approaches account for the influence of evolutionary forces, such as genetic drift or selection, on ancestry estimates (Jobling, Hurles, and Tyler-Smith 2013). Analytical methods underlying ancestry estimation can be categorized into two main approaches: those that are model based, or parametric, and those that are non-model-based, or non-parametric (Y. Liu et al. 2013). Parametric approaches rely upon specific assumptions about the characteristics of the genetic data in order to assess population substructure and assign individuals to subpopulations. Some of the assumptions for parametric approaches include that the markers of interest are in Hardy-Weinburg equilibrium and that genetic markers within each population are in linkage equilibrium, meaning that each genetic marker is independently inherited. In reality, these assumptions are often violated, and there are some statistical corrections that can be made to address these violations; however, these issues do not generally affect the utility of parametric approaches. Regardless, the underlying assumptions of parametric approaches are important to be aware of when interpreting analysis results (Kalinowski 2011). In general, parametric methods statistically infer population substructure and then assign individuals to the subpopulations. Non-parametric approaches do not rely upon assumptions about the genetic data and are computationally more efficient than parametric approaches. Non-parametric methods identify population substructure by grouping individuals based on genetic similarities or genetic distances and visualizing the clusters (Greenbaum, Templeton, and Bar-David 2016; Alhusain and Hafez 2018). The general idea in ancestry estimation is to systematically compare a test-taker's genetic markers to the substructured reference population. These comparisons can then be quantified and visualized to show the relationship between the test-taker and the reference population. The most reputable testing companies usually provide both summary and in-depth descriptions of how ancestry estimates are made (Durand et al. 2014; Ball et al. 2016; Curtis and Girshick 2017; Noto et al. 2018). Regardless of which analytical method is used, the quality of an ancestry estimation really hinges on the quality of the reference populations. As mentioned above, differences in reference populations including how reference populations are defined and the representativeness of reference groups can result in differences in ancestry estimations.

Types of ancestry tests and their limitations

Having reviewed what genetic ancestry tests are, the underlying conceptual framework, and analytical aspects of ancestry estimation, it is worth examining what genetic ancestry tests can actually say about a test-taker's ancestry. There are two broad groups of ancestry tests: uni-parental or unilineal ancestry tests and bi-parental or autosomal ancestry tests. Uni-parental tests provide ancestry information based on segments of the genome that are inherited from one parent and are therefore only informative about a single lineage within an individual's genetic genealogy. While the majority of our DNA is inherited from both parents, half from the mother and half from the father, two specific regions of the genome come from only one parent. These uni-parental segments of the genome are known as mitochondrial DNA (mtDNA) and Y-chromosome DNA.

MtDNA is a circular sequence consisting of approximately 16,569 base pairs and is found within a cellular organelle called mitochondria. Mitochondria are responsible for energy production within the cell and are in variable numbers depending on the cell type (Wallace 2007). Cells that require a lot of energy to function, such as brain, liver, or muscle cells, tend to have more mitochondria relative to other cell types (Cole 2016). While all people have mtDNA, only females pass it unchanged to offspring and only the female offspring can pass it to future generations. Because of this inheritance pattern, a person's mtDNA is the same as their mother's as well as their maternal grandmother, their maternal great-grandmother, and so on, back in time along the maternal lineage. Within mammals, the reasons for the maternal inheritance pattern of mtDNA from mother to child is unknown but is thought to occur in potentially different ways. One way that maternal mtDNA may be inherited is that because an ovum contains more mitochondria than a sperm, within a fertilized egg, the maternal mtDNA simply outnumbers the paternal mtDNA. Another explanation of the maternal inheritance of mtDNA includes the possibility of the active destruction of paternal mtDNA sometime during fertilization (Sato and Sato 2013). Given the maternal inheritance pattern, mtDNA is useful for gaining insights into the maternal lineage of an individual (Underhill and Kivisild 2007). However, because mtDNA is strictly maternally inherited, it is not particularly useful for gaining an understanding of a person's overall ancestry. In fact, mtDNA represents less than 1 percent of the DNA within a cell and consequently is not informative of the genetic contributions from relatives that are not along the maternal lineage.

Like mtDNA, the Y-chromosome is also only inherited from one parent. In this case, the Y-chromosome is inherited from the father by only his sons, and only males can pass it to future generations of males. A specific region of the Y-chromosome, known as the non-recombining region of the Y (NRY), contains the genetic information for making an individual chromosomally male. The NRY is what is used to assess paternal genetic ancestry. Since the Y-chromosome is inherited along the paternal lineage, it is only informative about the paternal lineage, meaning that it is not a useful part of the genome to use in learning about the general ancestry of any individual. Furthermore, for genetic ancestry test-takers, ancestry based on the Y-chromosome can only be assessed in people that are chromosomally male since individuals that are chromosomally female do not have Y-chromosomes. If a person that is chromosomally female wanted to learn about their paternal ancestry, they would need to test their father or a chromosomally male relative with a common male ancestor of their father, such as their father's brother or their brother.

Beyond being uni-parentally inherited, both mtDNA and Y-chromosome DNA contain genetic variants that are useful for assessing genetic ancestry. The genetic variants that occur within mtDNA and the Y-chromosome have been characterized and categorized into haplotypes. These haplotypes can then be grouped into haplogroups. There are over 5,000 mtDNA haplogroups (van Oven and Kayser 2009) and 311 Y-chromosome haplogroups (Zhang et al. 2013). The relationship between a haplotype and a haplogroup can be thought of in a similar way to a family. A family may contain siblings and these siblings will resemble one another due to shared parentage. However, there will be enough differences between the siblings that they are distinguishable from one another yet, there is enough similarity to be recognized as belonging to the same family. In the case of mtDNA and the Y-chromosome, haplotypes are the equivalent of siblings and haplogroups are the family. For each haplotype, there are slight differences between each respective haplotype; yet they are similar enough to be recognized as sharing a common origin, that is, the haplotypes are phylogenetically related and consequently grouped into the same family or haplogroup.

Haplogroups form as a result of random changes in the genetic sequence; this is known as genetic drift. They also can form in response to adaptation to local environments, otherwise known as natural selection (Mishmar et al. 2003). Both mtDNA and Y-chromosome haplogroups have been named after letters in the alphabet, where the first letter is the name of the haplogroup and subsequent letters and numbers identify sub-groupings under that haplogroup. The alphabetical naming scheme for uni-parental haplogroups is not indicative of genetic relationships between the haplogroups. Rather the haplogroups can be arranged in a tree diagram to illustrate the evolutionary relationships between the haplogroups. This tree diagram is known as a phylogenetic tree. Within academic studies, mtDNA and Y-chromosome haplotypes are useful for understanding phylogenetic relationships as well as for learning about the influence of various microevolutionary forces (e.g., genetic drift, gene flow, selection, and mutation) within the studied communities. In addition, mtDNA and Y-chromosome haplogroups are considered to be continent-specific, meaning that haplogroups are not uniformly distributed across the world, rather any given haplogroup tends to be more common in some parts of the world and less common in other parts of the world. The region in which any particular haplogroup is commonly found among indigenous populations is assumed to be around the geographic origin of that haplogroup. This, of course, assumes that the indigenous population has not migrated much from where they are currently found.

For the ancestry test- taker, the mtDNA and Y-chromosome haplogroups become important precisely because of the continental distribution of haplogroups. When a test-taker's mitochondria or Y-chromosome belongs to a particular haplogroup, their ancestry is inferred to come from wherever the haplogroup is thought to have originated. These types of inferences about ancestral origins are generally accepted and strengthened when there are other lines of evidence, such as archeological or linguistic data to support the purported ancestry. As an example, Y-chromosome haplogroup Q-M3 is commonly observed within indigenous populations in the Americas where nearly 75% of Native populations in Central and South America belong to Y-chromosome haplogroup Q (Bortolini et al. 2003; Battaglia et al. 2013). When a test-taker belongs to haplogroup Q-M3, they are said to have Native American ancestry along their paternal lineage. The possibility remains, however, that the test-taker's actual origin is from another region of the world where the haplogroup is found but is less common.

As previously mentioned, uni-parental haplogroups have extremely limited information content with regard to an individual test-taker's general ancestry. Uni-parental haplogroups are essentially only informative about one, or if chromosomally male, two, familial lineages. These types of ancestry tests provide no information about the vast majority of any person's general ancestry. Any relatives that fall outside of the direct maternal or paternal lineages, for example, a test-taker's mother's father or father's mother, will be invisible to this type of testing. In addition, as you go back in time, the number of your ancestors increases exponentially. You have two parents, your parents have two parents, each of your grandparents have two parents, and so on back each generation. If you do the math, in theory you have over one trillion ancestors 40 generations ago. That number is more than the number of people that exist today—current human population size is just over seven billion (http://www.worldometers. info/world-population/). In reality, you have far less than one trillion ancestors due to a phenomenon genealogists call "pedigree collapse" (Shoumatoff 1985; Cann 1988). This refers to people showing up multiple times within a pedigree and occurs with cousin marriages and the like. Though limited in scope, uni-parental tests provide a very small glimpse into a test-taker's genetic genealogy.

However, before discarding these types of tests as uninformative, it is worth noting that for some individuals, having some information, no matter how limited, is better than no information at all. In my work among various Caribbean communities, I have seen that for people who know nothing about their ancestry due to particular circumstances such as adoption, a general lack of knowledge about family history, or for people that are from communities that have been systematically denied accurate information about their ancestry, the glimpse into ancestral origins provided by these uni-parental haplogroups can be extremely valuable and meaningful in some very profound ways. Over the years, several people have contacted me after a study and shared personal stories about how their ancestry test results made them less scared or embarrassed to face some of the ugly facts of what it means to be a member of a community that was colonized or enslaved. As will be discussed throughout the remainder of the book and highlighted in the final chapter, though there is potential for ancestry tests to be used in manners that degrade at the dignity and sovereignty of communities, it is also important to recognize that genetic ancestry tests can also be wielded in ways that allow individuals to think about the past in different ways especially related to survival, resilience, and community continuity.

With regard to academic research, uni-parental ancestry tests are suited for investigating questions related to population genetics underlying human evolution as well as demographic histories. In this type of research, the combination of uni-parental ancestry from many individuals within a community or population provides insights into large-scale migration, gene exchange between populations, and relatedness within and between populations (Underhill and Kivisild 2007; Bryc et al. 2015; Nielsen et al. 2017). Due to the utility of uni-parental ancestry tests for learning information, though limited, about demographic and genealogical histories, these types of ancestry tests will remain a mainstay within the DTC industry as well as in genetic anthropological research.

The second broad category of ancestry tests are known as bi-parental or autosomal ancestry tests. Autosomal tests are used in local and global ancestry estimation. Unlike mtDNA and the Y-chromosome, autosomal DNA has a more complex inheritance pattern, which has the end result of providing a more comprehensive picture of a test-taker's overall genetic ancestry. This portion of the genome is inherited from both parents, where each parent contributes half of their DNA to the offspring. In addition, also unlike the uni-parentally inherited DNA, autosomal DNA recombines, meaning that the DNA is shuffled within each parent during the process of sex cell production. Essentially, this process of recombination has the effect of making an individual a mosaic of their ancestors' genetic contributions. Within the context of genetic ancestry tests, both the bi-parental inheritance and the mosaic nature of autosomal DNA is what allows for the generalized assessment of a test-taker's ancestry across all genealogical lineages. As discussed above, ancestry tests using autosomal DNA rely upon comparisons of AIMs between the test-taker and reference populations. Consequently, a major limitation of autosomal ancestry tests is that the quality of any ancestry test is heavily reliant on the comprehensiveness of included AIMs and reference populations. In addition, as mentioned in our discussion of local and global ancestry, the genetic ancestry results provided to a test-taker have the potential to change when the AIMs or reference populations are modified because the comparisons between the test-taker and reference populations become different. Differences in which AIMs are used and which reference populations are included in ancestry analyses explains why consumers can get different results when testing with different ancestry companies or when a company updates its reference populations. For the DTC ancestry test-takers, the more reputable testing companies do provide information about the composition and extent of their reference populations as well as some indication of the number of genetic markers, including AIMs, used in the ancestry tests.

As a final note, as researchers continue to make improvements to the analytical and methodological aspects of genetic ancestry tests, it is nonetheless crucial to remember that genetic data are best understood when put into appropriate historical and social contexts. We refer to contextualizing genetic data as gaining some understanding of the circumstances that may have contributed to shaping genetic diversity. For the individual test-taker, this could mean having some idea about family or community history as this can help to make sense of ancestry results. Within academia, adding context can include drawing evidence from multiple sources, such as archeology or archival and oral histories, and using that information as lenses to interpret the genetic data. While genetic ancestry estimates are often presented as seemingly objective concise quantities, we end this chapter making the point that it is important for any person engaged in ancestry testing to remember that ancestry estimates are a product of particular ways of understanding, querying, and organizing the world. Here, we want to emphasize that ancestry estimates are not created in a vacuum but instead are reflective of particular histories and ways of thinking about and engaging with the past. In the next chapter, we expand on the significance of contextualizing genetic ancestry findings. In our discussion of this, we examine the philosophical and anthropological underpinnings of how people think about the past and how understanding of the past shapes how people engage with the present, including their genetic data.

References

Agro, Charlsie, and Luke Denne. 2019. "Twins Get Some 'Mystifying' Results When They Put 5 DNA Ancestry Kits to the Test." *CBC News*, January 18, 2019. https://www.cbc.ca/news/technology/dna-ancestry-kits-twins-marketplace-1.4980976.

Alhusain, Luluah, and Alaaeldin M. Hafez. 2018. "Nonparametric Approaches for Population Structure Analysis." *Human Genomics* 12 (1): 25. doi:10.1186/s40246-018-0156-4.

Ball, Catherine A., Mathew J. Barber, Jake Byrnes, Peter Carbonetto, Kenneth G. Chahine, Ross E. Curtis, Julie M. Granka, et al. 2016. *Ancestry DNA Matching White Paper.*

Battaglia, Vincenza, Viola Grugni, Ugo Alessandro Perego, Norman Angerhofer, J. Edgar Gomez-Palmieri, Scott Ray Woodward, Alessandro Achilli, et al. 2013. "The First Peopling of South America: New Evidence from Y-Chromosome Haplogroup Q." *PLOS One* 8: e71390.

Borry, Pascal, Martina C. Cornel, and Heidi C. Howard. 2010. "Where Are You Going, Where Have You Been: A Recent History of the Direct-to-Consumer Genetic Testing Market." *Journal of Community Genetics* 1 (3): 101–6. doi:10.1007/s12687-010-0023-z.

Bortolini, M. C., F. M. Salzano, M. G. Thomas, S. Stuart, S. P. Nasanen, C. H. Bau, M. H. Hutz, et al. 2003. "Y-Chromosome Evidence for Differing Ancient Demographic Histories in the Americas." *American Journal of Human Genetics* 73: 524–39. doi:10.1086/377588.

Bryc, K., E. Y. Durand, J. M. Macpherson, D. Reich, and J. L. Mountain. 2015. "The Genetic Ancestry of African Americans, Latinos, and European Americans across the United States." *American Journal of Human Genetics* 96: 37–53. doi:10.1016/j.ajhg.2014.11.010.

Cann, R. L. 1988. "DNA and Human Origins." *Annual Review of Anthropology* 17 (1): 127–43. doi:10.1146/annurev.an.17.100188.001015.

Cole, Logan W. 2016. "The Evolution of Per-Cell Organelle Number." *Frontiers in Cell and Developmental Biology* 4 (August): 85. doi:10.3389/fcell.2016.00085.

Curtis, Ross E., and Ahna R. Girshick. 2017. "Estimation of Recent Ancestral Origins of Individuals on a Large Scale." In *Proceedings of the 23rd ACM SIGKDD International Conference on Knowledge Discovery and Data Mining*, 1417–425. KDD '17. New York: ACM. doi:10.1145/3097983.3098042.

Durand, Eric Y., Chuong B. Do, Joanna L. Mountain, and J. Michael Macpherson. 2014. "Ancestry Composition: A Novel, Efficient Pipeline for Ancestry Deconvolution." *bioRxiv*, 010512.

Fan, Shaohua, Matthew E. B. Hansen, Yancy Lo, and Sarah A. Tishkoff. 2016. "Going Global by Adapting Local: A Review of Recent Human Adaptation." *Science* 354: 54–59. doi:10.1126/science.aaf5098.

Greenbaum, Gili, Alan R. Templeton, and Shirli Bar-David. 2016. "Inference and Analysis of Population Structure Using Genetic Data and Network Theory." *Genetics* 202 (4): 1299–312. doi:10.1534/genetics.115.182626.

Hodgson, Jason A., Joseph K. Pickrell, Laurel N. Pearson, Ellen E. Quillen, António Prista, Jorge Rocha, Himla Soodyall, et al. 2014. "Natural Selection for the Duffy-Null Allele in the Recently Admixed People of Madagascar." *Proceedings of the Royal Society B: Biological Sciences* 281 (1789): 20140930. doi:10.1098/rspb.2014.0930.

Hoggart, C. J., E. J. Parra, M. D. Shriver, C. Bonilla, R. A. Kittles, D. G. Clayton, and P. M. McKeigue. 2003. "Control of Confounding of Genetic Associations in Stratified Populations." *American Journal of Human Genetics* 72: 1492–504.

Jobling, Mark, Matthew Hurles, and Chris Tyler-Smith. 2013. *Human Evolutionary Genetics: Origins, Peoples & Disease*. New York. Garland Science.

Jobling, Mark A., Rita Rasteiro, and Jon H. Wetton. 2016. "In the Blood: The Myth and Reality of Genetic Markers of Identity." *Ethnic and Racial Studies* 39 (2): 142–61. doi:10.1080/01419870.2016.1105990.

Kalinowski, S. T. 2011. "The Computer Program STRUCTURE Does Not Reliably Identify the Main Genetic Clusters within Species: Simulations and Implications for Human Population Structure." *Heredity* 106 (4): 625–32. doi:10.1038/hdy.2010.95.

Kano, Flora Satiko, Aracele Maria de Souza, Leticia de Menezes Torres, Marcelo Azevedo Costa, Flávia Alessandra Souza-Silva, Bruno Antônio Marinho Sanchez, Cor Jesus Fernandes Fontes, et al. 2018. "Susceptibility to Plasmodium Vivax Malaria Associated with DARC (Duffy Antigen) Polymorphisms Is Influenced by the Time of Exposure to Malaria." *Scientific Reports* 8 (1): 13851. doi:10.1038/s41598-018-32254-z.

Kwiatkowski, Dominic P. 2005. "How Malaria Has Affected the Human Genome and What Human Genetics Can Teach Us about Malaria." *The American Journal of Human Genetics* 77 (2): 171–92. doi:10.1086/432519.

Lachance, Christina R., Lori A. H. Erby, Beth M. Ford, Vincent C. Allen Jr., and Kimberly A. Kaphingst. 2010. "Informational Content, Literacy Demands, and Usability of Websites Offering Health-Related Genetic Tests Directly to Consumers." *Genetics in Medicine* 12 (5): 304–12. doi:10.1097/GIM.0b013e3181dbd8b2.

Leighton, J. W., K. Valverde, and B. A. Bernhardt. 2012. "The General Public's Understanding and Perception of Direct-to-Consumer Genetic Test Results." *Public Health Genomics* 15 (1): 11–21. doi:10.1159/000327159.

Leslie, Stephen, Bruce Winney, Garrett Hellenthal, Dan Davison, Abdelhamid Boumertit, Tammy Day, Katarzyna Hutnik, et al. 2015. "The Fine-Scale Genetic Structure of the British Population." *Nature* 519 (7543): 309–14. doi:10.1038/nature14230.

Liu, Jinghua, Juan Pablo Lewinger, Frank D. Gilliland, W. James Gauderman, and David V. Conti. 2013. "Confounding and Heterogeneity in Genetic Association Studies with Admixed Populations." *American Journal of Epidemiology* 177: 351–60. doi:10.1093/aje/kws234.

Liu, Yushi, Toru Nyunoya, Shuguang Leng, Steven A. Belinsky, Yohannes Tesfaigzi, and Shannon Bruse. 2013. "Softwares and Methods for Estimating Genetic Ancestry in Human Populations." *Human Genomics* 7 (1): 1.

Luzzatto, L. 2012. "G6PD Deficiency and Malaria Selection." *Heredity* 108: 456. doi:10.1038/hdy.2011.90.

Mackinnon, Margaret J., Carolyne Ndila, Sophie Uyoga, Alex Macharia, Robert W. Snow, Gavin Band, Anna Rautanen, et al. 2016. "Environmental

Correlation Analysis for Genes Associated with Protection against Malaria." *Molecular Biology and Evolution* 33 (5): 1188–204. doi:10.1093/molbev/msw004.

Mishmar, D., E. Ruiz-Pesini, P. Golik, V. Macaulay, A. G. Clark, S. Hosseini, M. Brandon, et al. 2003. "Natural Selection Shaped Regional MtDNA Variation in Humans." *Proceedings of the National Academy of Sciences of the United States of America* 100: 171–76.

Nielsen, Rasmus, Joshua M. Akey, Mattias Jakobsson, Jonathan K. Pritchard, Sarah Tishkoff, and Eske Willerslev. 2017. "Tracing the Peopling of the World through Genomics." *Nature* 541 (7637): 302–10. doi:10.1038/nature21347.

Norrgard, K. 2008. "DTC Genetic Testing for Diabetes, Breast Cancer, Heart Disease and Paternity." *Nature Education* 1 (1): 86.

Noto, Keith, Yong Wang, Shiya Song, David Turissini, Alisa Sedghifar, Barry Starr, Jake Byrnes, et al. 2018. "Ethnicity Estimate 2018 White Paper." November 30, 2018. https://www.ancestrycdn.com/dna/static/pdf/whitepapers/WhitePaper_2018_1130_update.pdf.

Oven, M. van, and M. Kayser. 2009. "Updated Comprehensive Phylogenetic Tree of Global Human Mitochondrial DNA Variation." *Human Mutation* 30: e386–94.

Pearson, Yvette E., and Yuping Liu-Thompkins. 2012. "Consuming Direct-to-Consumer Genetic Tests: The Role of Genetic Literacy and Knowledge Calibration." *Journal of Public Policy & Marketing* 31 (1): 42–57. http://www.jstor.org/stable/41714254.

Phillips, Andelka M. 2016. "'Only a Click Away—DTC Genetics for Ancestry, Health, Love…and More: A View of the Business and Regulatory Landscape.'" *Applied & Translational Genomics*, Personal Genomics: Complications and Aspirations, 8 (March): 16–22. doi:10.1016/j.atg.2016.01.001.

Price, Alkes L., Noah A. Zaitlen, David Reich, and Nick Patterson. 2010. "New Approaches to Population Stratification in Genome-Wide Association Studies." *Nature Reviews Genetics* 11 (7): 459–63. doi:10.1038/nrg2813.

Regalado, Antonio. 2018. "2017 Was the Year Consumer DNA Testing Blew Up." *MIT Technology Review*, February 12, 2018. https://www.technologyreview.com/s/610233/2017-was-the-year-consumer-dna-testing-blew-up/.

———. 2019. "More than 26 Million People Have Taken an At-Home Ancestry Test." *MIT Technology Review*, February 11, 2019. https://www.technologyreview.com/s/612880/more-than-26-million-people-have-taken-an-at-home-ancestry-test/.

Rosen, Christine. 2003. "Liberty, Privacy, and DNA Databases." *The New Atlantis* 1: 37–52. https://www.jstor.org/stable/43152851.

Rosenberg, N. A., L. M. Li, R. Ward, and J. K. Pritchard. 2003. "Informativeness of Genetic Markers for Inference of Ancestry." *American Journal of Human Genetics* 73: 1402–422.

Sato, Miyuki, and Ken Sato. 2013. "Maternal Inheritance of Mitochondrial DNA by Diverse Mechanisms to Eliminate Paternal Mitochondrial DNA."

Biochimica et Biophysica Acta (BBA) – Molecular Cell Research 1833 (8): 1979–984. doi:10.1016/j.bbamcr.2013.03.010.

Saukko, Paula. 2013. "State of Play in Direct-to-Consumer Genetic Testing for Lifestyle-Related Diseases: Market, Marketing Content, User Experiences and Regulation." *The Proceedings of the Nutrition Society* 72 (1): 53–60. doi:10.1017/S0029665112002960.

Shoumatoff, Alex. 1985. "A Reporter at Large: The Mountain of Names." *The New Yorker* 51.

Statista. 2017. "Direct-to-Consumer Genetic Testing Market Size Worldwide 2014–2022 | Statistic." *Statista*, December 2017. https://www.statista.com/statistics/792022/global-direct-to-consumer-genetic-testing-market-size/.

Su, Pascal. 2013. "Direct-to-Consumer Genetic Testing: A Comprehensive View." *The Yale Journal of Biology and Medicine* 86 (3): 359–65. https://www.ncbi.nlm.nih.gov/pmc/articles/PMC3767220/.

Thornton, Timothy A., and Justo Lorenzo Bermejo. 2014. "Local and Global Ancestry Inference, and Applications to Genetic Association Analysis for Admixed Populations." *Genetic Epidemiology* 38 (01): S5–12. doi:10.1002/gepi.21819.

Underhill, Peter A., and Toomas Kivisild. 2007. "Use of Y Chromosomes and Mitochondrial DNA Population Structure in Tracing Human Migrations." *Annual Review of Genetics* 41: 539–64.

Wallace, Douglas C. 2007. "Why Do We Still Have a Maternally Inherited Mitochondrial DNA? Insights from Evolutionary Medicine." *Annual Review of Biochemistry* 76 (1): 781–821. doi:10.1146/annurev.biochem.76.081205.150955.

Zhang, Fan, Ruoyan Chen, Dongbing Liu, Xiaotian Yao, Guoqing Li, Yabin Jin, Chang Yu, et al. 2013. "YHap: Software for Probabilistic Assignment of Y Haplogroups from Population Re-Sequencing Data." *ArXiv:1304.3351 [q-Bio]*, April. http://arxiv.org/abs/1304.3351.

2 How do humans think about the past and their ancestors?

"You will see that I am a son of Joseph Chatoyer!" This is what, Delroy, a young Garifuna man loudly declared after volunteering his genetic sample to be included in my genetic ancestry study of St. Vincent's Garifuna community. I (Jada) listened to Delroy intently as he left the makeshift sample collection site, which, on most days, was a seldom used storage office. We were working near the center of Rose Bank, a rural community on the west coast of St. Vincent and just north of the island nation's capital, Kingstown. Vincentians recognize this community as home to people of mixed Indigenous Caribbean and African ancestry, more commonly known as Garifuna. As I carefully organized Delroy's sample materials into my fieldwork bag, he walked across the adjacent yard and, with youthful but innocent pride, pounded his chest while asserting his connection to Paramount Chief Joseph Chatoyer. Chatoyer, St. Vincent's national hero, was a fierce eighteenth-century Garifuna resistance fighter and contemporary symbol of Vincentian pride. Delroy's claim to Chatoyer was his way of illustrating his membership firmly in the Garifuna community and, probably more directly, in response to the genealogical interviews I had completed with him during his enrollment in my project.

This scene in St. Vincent was one of many instances in which people from throughout the Caribbean interacted with us for the purposes of genetic ancestry research. Sometimes our conversations with research participants were very formal, almost businesslike. Other times, it was like collecting samples from friends, and it was very much a familial atmosphere. Regardless of the setting, it became very clear to us that we could not take for granted what ideas people have about our research and genetic ancestry in general. Most importantly, we learned that people already have ideas about their ancestors and their communities. When researchers like us show up looking to set up a project, recruit participants, and return results, these existing ideas about ancestors and communities affect how participants interpret all aspects of genetic ancestry research. For some, genetic ancestry is irrelevant to how they know the past. For others, it becomes a focal point of discovering their past. For many, genetic ancestry is a curiosity that, although technically difficult to understand, they are willing to try out. Whatever the case, paying attention to these notions of ancestors and communities allows us to better understand how and why Delroy spoke with such pride and certainty about being a Garifuna man and specifically a descendent of Joseph Chatoyer—and why the genetic ancestry results would demonstrate what he said he already knew.

The previous chapter explained how genetic ancestry can provide insights into our ancestral past, so in this chapter we examine some important ways in which humans think about ancestry. Given the vast range of human creativity over time and across the world, it is impossible to provide a comprehensive overview of human cultural ancestry (indeed, anthropologists and others have filled volumes on this topic). However, we do think it is possible to present some of the most important social and cultural dimensions of how humans think about *the past and their ancestors* (this chapter) that, in turn, affect how they receive and interpret genetic ancestry information (next chapter).

Ancestry, from a social and cultural perspective, has two important dimensions: the past and relatedness. That is, our ancestors are relatives in our past or those that have been in the world before we came into the world. This chapter will first take up some philosophical considerations about how "the past" is an aspect of time that humans understand in relation to the spaces they inhabit, their understanding of the present and future, and their consciousness. This brief philosophical consideration outlines the possibilities and limitations of how humans imagine the past and sets up an explanation of an anthropological approach to cultural ancestry. Having considered philosophical and anthropological dimensions of the past, we move to a discussion of relatedness, which is important for understanding who is included and of significance when humans think about their ancestors. Finally, we consider the linguistic importance of discourses about ancestry since when people talk about their ancestors they are both saying something about the past and what they consider important in the present.

Time, space, and human consciousness

The past is an aspect of time, and scholars working in diverse disciplines have examined time from multiple dimensions (see Callender 2011b;

Bouton and Huneman 2017, for example, for comprehensive efforts to summarize the study of time). Questions about time look different in the study of, say, quantum mechanics, relativistic physics, paleontology, ethics, metaphysics, and cognitive science. We have to put aside many of these scholarly takes on time because there is too much material to cover and most of it falls outside the scope of this book; but it is important to highlight two interrelated considerations when thinking about time: space and relativity.

Space is important because, like time, both are parts "of the fundamental stage upon which the events of the world play out" (Callender 2011a, 1). Simply stated, time presumably unfolds in space, but this presumption is itself problematic. For most (if not all) humans, having a full account of all the space in the universe is impossible, so people have forged various notions of space in finite forms. Religions and astrophysics, for example, often strive to formally outline the boundaries of known worlds or universes. There are also more quotidian ways to delimit space. For example, we learn and adopt particular ways to move from one place to another. We say we are from "this city" or "this country" even though there are parts of our cities we rarely visit, and most people certainly have neither visited nor are spatially aware of most of "their country." We may know we live on a planet in the universe, but we are likely to only occupy an infinitesimal fraction of our planet throughout our lives. For humans, then, the notion of space in which time unfolds varies; and since cultural notions of space vary, cultural notions of time also vary. A simple example from personal experience comes from when I (Gabriel) was growing up in Puerto Rico. On occasion, I would take a roughly 70 mile-road trip from the south to the north of the island. I and many Puerto Ricans understood this journey as being a very long trip. This is underscored by the fact that there is an actual highway rest stop halfway through the route. However, like many Puerto Ricans and other people that grew up in small islands, when I moved to the United States—where commonly traveled distances are often much longer—the perception of that island trip as "long" changed.

The relationship between time and space is, in fact, one way in which time is relative. Time is relative in other ways, take for example, Albert Einstein's special theory of relativity, which explains how time is different for objects moving at different speeds (for a general primer, see Egdall 2014). In philosophical metaphysics, there is a long-standing debate regarding the nature of three dimensions of time: past, present, and future. Philosophers debate (1) the reality of past and future relative to the present, and (2) they consider the nature of things, particularly humans, as they move through time. Let us consider each of these debates in order.

First, with regards to *the reality of past and future relative to the present*, there are different schools of thought. Some argue that past, present, and future are all real; yet others argue that only the present is real and that the past and the future can only be imagined in the present (Callender 2011a, 3). It is important to not confuse these philosophical positions with scientific and non-scientific approaches to the past, present, and future since scientists themselves have led the way in dismantling so-called commonsense approaches to the reality of all time (Anderson 2017). Of course, there are many scientific fields, such as evolutionary biology, that assume that the scientific methods can help us reach empirical (real) conclusions about past events. Regardless of where one stands philosophically with regard to the reality of time, what is most important for this chapter is that the past is conceptually tied to the present and the future. Therefore, if one believes that all time is real, then how we understand the past is dependent on our empirical possibilities for probing the past in the present. If one believes the past is not real, then the past is something that we only experience as real in the present. We actually believe that, when it comes to human affairs, both of these positions are compatible; but we do not need to resolve this issue right now. We will return to this debate at the end of the chapter.

Second, another philosophical debate is concerned with *the nature of things as they move through time*, particularly humans. Along with many other life forms, humans change within their life span. Humans also experience change between generations, and they ultimately experience change over the evolutionary history of their species. The key question here is to what degree we, in the present, constitute what we have been and what we will become. Again, like the question about time and space, we do not need to resolve the nature of "being" through time, but what we do need to understand is that humans cannot possibly account for every instance of their being—past or future—in an all-encompassing way. Some of these instances are easy to understand, such as when you sleep or before you have awareness of yourself as an infant. Yet there are other instances of being that we might think of as unmemorable—for example, an uneventful walk to work or making yourself breakfast sometime three years ago. These instances fill our lives, yet they do not tend to be part of our consciousness and, consequently, the stories that we tell about ourselves through time. William James has famously explained,

Consciousness, then, does not appear to itself chopped up in bits. Such words as "chain" or "train" do not describe it fitly as it presents itself in the first instance. It is nothing joined; it flows. A "river" or a "stream" are the metaphors by which it is most naturally described.

(James 2007, 239)

Therefore, both our notion of time and our notion of ourselves through time are dependent on the nature of our consciousness. Note that we are not about making the claim that time is completely relative to human consciousness; instead, we are claiming that our ability to perceive time and ourselves in time is dependent on human consciousness.

Let us summarize the philosophical insights on time so far. The past is an aspect of time. Time unfolds in space, and therefore time is relative to space. In addition, time is relative to itself (past, present, and future) and relative to how humans change through time and perceive their perception of this change. We can now consider these insights from an anthropological point of view in order to understand how genetic ancestry is one of many ways in which humans can think about themselves and the past.

Time from an anthropological perspective

Since anthropology is concerned with human diversity, we might begin by questioning the universality of some of the philosophical concerns we have just discussed. Time itself can be studied from multiple anthropological angles. Cultural and linguistic anthropologists have asked if people everywhere in the world have similar conceptions and ways of speaking about time. The length of existence of the world, for example, varies on religious worldviews and other ways of thinking, like evolutionary science. Languages also vary significantly in allowing speakers to denote past events (i.e., past tenses). In the interdisciplinary field of neuroanthropology (Lende and Downey 2012), one might question how these different linguistic conceptions of time and time frames of lived experience differently shape neuro-cognitive processes (Lucy 1996). In fact, there are many known cultural variations of how humans think about and experience time, and the important point of these anthropological approaches to time is that we cannot assume that concepts of time play the exact same role in every culture. Yet knowing that there is cultural variation should not necessarily lead us to the idea that time is only real in relation to how humans in different cultures understand it. As Alfred Gell (1992) argues, the anthropology of time can both question the multifold ways in which humans frame time and hold on to the assumption that time is universal. Or to put it differently, we can assume that time occurs independent of human experience, and we can also question how humans differently think about and adapt to time.

We can similarly question space as it is differently conceived in different cultures around the world and throughout history and prehistory. In fact, Keith Basso (1996) has argued that rather than thinking that humans differently inhabit the same type of objective space, we should conceive of humans as constantly engaged in processes of place-making. Like time, humans do not really experience objective abstract spaces outside of their culturally meaningful worlds. Instead, places are evocative of symbolic meaning that reveal aspects of our cultures. For example, cemeteries reveal religious worldviews, monuments reflect nationalist sentiments, and a public park might relieve us from the stress of everyday work and family routines. Even when thinking of an empty space, humans have a wide range of perspectives, from peaceful emptiness to scary and unfamiliar—thoughts that actually have meaning. Therefore, humans might be delimited in their abilities to fully comprehend or imagine every aspect of space external to their perception, yet in every sociocultural group we can observe people meaningfully imbuing the spatial world with meaning.

These general anthropological lessons about time and space/place are important when thinking about ancestry. Ancestry, whether genetic or cultural, is about humans in past places that we consider meaningful to our lives; yet genetic ancestry is one of many ways in which humans think about the past. In fact, everyone who encounters genetic ancestry already has existing cultural forms that shape how they think about the past. The question remains of how significant is genetic ancestry in shaping these existing ideas about the past. The answer, which we will return to at the end of this chapter, is contingent on cultural notions of relatedness. Genetic ancestry information cannot only be understood through its time dimension because, if it was, genetic ancestry companies would not have much to offer other than confirming that all humans come from a long line of humans that dates back to the beginning of our species 200,000 to 300,000 years ago. Instead, people have working understandings of the people to whom they are related, so an individual's inquiry into her past is also an inquiry into the people who she considers her people. Of course, how people think about "their people," or relatedness, varies cross-culturally.

Relatedness

Who are your ancestors? The answer, or potential multiple answers, will immediately say something about who you think is related to you in the past—those that came before you and led to you. Your answer might also say something important about the present since you will share your ancestors with groups of people who are contemporaneous with you. Any one individual is part of various social groups that exist in different social spaces; moreover, these social groups will have different historical timelines. For example, your family, your hometown, your sorority, your gender, your religion, and your nationality might all tell slightly different versions of who you are and who your people are. Anthropologists have examined different aspects of this social complexity by studying the structures of relatedness through which humans engage in social behaviors. There are at least two dimensions to these structures of relatedness: social and symbolic. By social structures we mean the patterns of the behavioral interactions between people, and by symbolic structures we mean the collectively shared ideas used to represent social groups that people belong to.

There is a lot of overlap between social and symbolic structures, although there can be some differences. For example, if we think about "class," it can refer to a group of people who tend to live in similar areas, earn similar incomes, have similar education levels, and occupy similar social spaces. There are interactional structures that we can empirically assess for this hypothetical group of people. However, the symbolic structure of "class" can include different social groups living in different places, earning different incomes, and with different educational levels-such is the case for people who claim they are part of the "middle class" in the United States. This example illustrates that, the concept of "middle class" can both refer to a social structure of, say, the hypothetical suburban neighborhood composed of similar income earners, and the middle class can be a symbolic structure that people from many social, economic, and political backgrounds identify with. The anthropological concept of "culture," in its most holistic articulation, can be theoretically helpful in simultaneously thinking about social and symbolic structures (Sahlins 2000, 16) by allowing us to probe how and to what degree human experience is shaped and shapes social and symbolic structures (Rodseth 1998). When we then ask what are the cultural dimensions of ancestry, we are asking questions about both the social and symbolic structures of people's ancestry that include relatedness to others, living and dead, and also disassociation with others; yet these structures are complex and never straightforward with the overlapping and shifting notions of the stories that people tell about themselves and their pasts.

Through years of fieldwork collecting samples, conducting interviews, and analyzing data, there are three cultural dimensions of ancestry that we think are important to understanding questions about relatedness, and ancestry more specifically: kinship, nationalism, and embodied difference. Each one of these cultural dimensions has social and symbolic structures, and depending on context, often overlap.

Kinship

In anthropology, the study of kinship is as old as the discipline itself, dating back to the late nineteenth century. Several important schools of thought developed over the twentieth century. Early and mid-twentieth century traditions focused on the importance of kinship systems in non-Western societies. For some anthropologists, kinship was the glue that held together and established order in societies that did not have formal institutions of governance; while other anthropologists focused on the role of symbolic and linguistic structures of kinship in regulating economic exchange and social relations (Carsten 2003, 10–16). Late twentieth century traditions criticized the rigidity of earlier paradigms and emphasized the symbolic flexibility and creativity through which people meaningfully relate to one another (Schneider 1980; Peletz 1995; Parkin and Stone 2004). Contemporary kinship theories tend to emphasize a notion of "relatedness" that can account for multiple dimensions of social and symbolic structures. We want to highlight some of these dimensions; although a comprehensive review is not possible, a brief discussion will allow enough grounding to contextualize our analysis of genetic ancestry.

First, in all human societies, there is a cultural concern with the relationships that form around processes of reproduction. Here, reproduction is not only concerned with birth, but also mating and child rearing. This definition of reproduction takes into account the fact that humans are, evolutionarily speaking, a social species. Therefore, reproduction cannot be reduced to the basic biology of reproduction and must take into account the social mechanisms through which human species survival is made possible. Sexuality, birthing, and child development are both biological and social in nature.

However, because cultural anthropologists have documented many instances in which genetic relationships do not explain a kinship system, they have rejected the idea that biology and kinship are intrinsically linked. We think that this assertion is well-intentioned but fundamentally flawed—not because there is a strict relationship between genetic descent and kinship—but because so-called non-biological acts such as child rearing and non-reproductive sex are, in fact, biological. For example, an adopted baby can receive nurturance from his parents that has biological effects for both child and parent (e.g., Gettler and McKenna 2011; Gettler et al. 2011). We understand and acknowledge anthropological concerns with countering the reduction of kinship to biology because there are too many instances around the world when the equivalence of biology and kinship are used as a justification for establishing what sorts of relationships are "normal" or "morally right." We agree with those pedagogical efforts, but we also do not want to sacrifice substance in light of complexity. Genetic relationships do not necessarily lead to certain kinship forms, but genetic relationships are almost always symbolically represented through kinship forms. All of the major case studies in this book, for example, involve societies in which genetic descent is recognized in some form as being relevant to kinship systems. What we must recognize, then, is that where we see people thinking about how they are related to one another and taking into consideration who gave birth to who, we cannot come to the conclusion that the kinship system is *intrinsically* linked to genetic descent; that is reproduction is consistently important to understanding kinship, but reproduction is not the only thing that matters or necessarily the most important factor in determining relatedness.

Second, kinship systems are ways in which people engage in what Marshall Sahlins calls "mutuality of being" or "people who are intrinsic to one another's existence" (Sahlins 2013, 2). This mutuality of being is discernible through patterns of familial interactions and the meaning attached to those interactions. Kinship terminologies—the terms that people use to identify family (or clan, tribes, etc.)—are often reflective of these interactions. Examples of kinship terms include mother father, stepsister, aunt, and others; and the anthropological insight here is that these terms vary across cultures because how people determine relatedness also varies. Of course, the kinship terms are not mere reflections of relationships, often humans use language to assign meaning to relationships (e.g., sorority members calling each other "sisters"). This dimension of kinship is very complex, and cultural anthropologists have dedicated volumes to fleshing out the dimensions and possibilities of how humans meaningfully relate to one another. One important aspect of this vast body of research is the notion that kinship unfolds within systems of power. This means that kinships systems can lend themselves to the upholding of hierarchies (e.g., patriarchies) or to the exclusion and alienating of certain kin relations (e.g., the unlawfulness of gay marriage).

Third, kinship systems also serve as metaphors for other forms of social relationships. For example, a small town might be made up of a dozen families, and people in that town might see themselves as a small "family town." A racial group in a country might live in segregation, and the members of that racial group might use kinship terms—such as "brother" or "sister"—to refer to other members of their racial group. Or officers in a police department might have a long history of members coming from certain families in town, and the officers in that department might see themselves as a "fraternal order." We can consider the articulation of this dimension of kinship as a way of acknowledging that social systems exist on different scales and with overlapping complexities; but it is also a way of acknowledging that models for what constitute "a family" influence and are influenced by broader social and symbolic structures.

Nationalism

Nations are imagined (Anderson 2006), represented (Kelly 2001), and built over time (Corrigan and Sayer 1991). Nations are imagined because citizens cannot possibly have a social relationship with almost the entire population; and people who identify with the same nationality are also socially diverse: from different classes, ethnicities, and linguistic groups. Therefore, their common nationality is a cultural phenomenon that exists to alleviate that difference—and sometimes at the expense of those who do not want to be alleviated of their difference. Nations are represented for a similar reason. Any nation has had historically competing versions of "the nation," so people must articulate their vision of the nation; and doing so can take many forms: from writing history books, to political speeches, to the desecration of a national monument. Finally, nations are built over time because the imagination and representation of a nation is a historical process that involves wars, peace treaties, building of cities, environmental catastrophes, technological advances, and many other occurrences that shape human societies.

The "history" of a nation is one of the most important narratives throughout which people represent their nation and the nations of others. National histories bring order to messy pasts and excluded alternative versions of history. Furthermore, national histories are often powerful because they are part of socialization processes, particularly through schooling, public ceremonies, and marketed forces.

Consequently, how people think about their ancestry is intricately tied to ideas about the nation. People suppress, overemphasize, question, or ignore their ancestry in relation to national narratives. For example, a nationalist narrative can exclude people with certain ancestry and define a nation through a particular set of ancestors. Finally, leaders of nations and nations themselves take on ancestral metaphors to define themselves. As mentioned above, kinship terminology—like motherland or fatherland—become synonymous with a nation, and "founding fathers" can be a way of describing the ancestral origins of a nation. For the vast majority of people who live within the bounds of a country, nationalism plays a significant role in either incorporating or excluding other sub-national and local forms of social identification. In many ways, nationalism takes up the social and symbolic structures of kinship systems, and expands these structures in the interest of expanding the power of those social sectors that already hold power in a country.

Embodied difference

Embodied difference, which we have elsewhere termed "bioculturally embodied difference," is a way to operationalize the concept of "race" (Benn Torres and Torres Colón 2016; Torres Colón 2018) and refers to the gendered and racial social groups that emerged within and across nations. Embodied difference creates symbolic structures whereby groups of people experience the social world through differences that are embodied—that is, traits that are seen as inherent to certain types of bodies. Furthermore, those embodied differences are part of symbolic structures that order the world into differentiated social bodies. The nature of the differentiation can be either physical (bodies), psychical (minds), or both. White women, black men, white men, black women, indigenous women, and indigenous men are concrete examples of embodied difference in the Americas. Embodied differences do not exist in socially distinct groups; instead, there are overlapping groupings and subgroupings. For example, indigenous women and men, obviously, often exist in the same social space; yet the symbolic structures attached to their embodied difference might vary (Nagel 2003; Hill Collins 2004).

There are significant overlaps between structures of embodied difference and structures of kinship and nationalism. For example, in Latin American *mestizaje* ("mixture") can serve as a model for the nation, and as we mentioned above, kinship ideas can shape the way gender and racial groups see themselves. However, unlike kinship and nationalism, embodied differences fully merges bodies with the experiences of those bodies as differentiated bodies. Whereas bodies of families are often tied through close genetic relationships, embodied difference ties bodies of social groups well beyond immediate genetic relationships. That tie is based on likeness of bodies and the experience of those bodies as different. These experiences of embodied difference can be charted along a broad spectrum of emotions, ranging from positive (e.g., racial pride), to neutral, to negative (e.g., racism).

We are aware that most readers will be encountering the idea of embodied difference as a form of relatedness here for the first time—especially in juxtaposition to kinship and nationalism. Unlike kinship and nationalism, embodied difference does not appear in introductory anthropology texts, although sections about race and gender do. We ask the reader to entertain embodied difference as a core theoretical concept because, after years of fieldwork with racialized communities in the Caribbean and the United States, race—as an expression of embodied difference—has appeared as a distinct prism through which people think about their ancestors and genetic ancestry. We believe that privileging a racialized perspective in a discussion of how humans think about relatedness and the past is how most anthropologists should think about such issues, but this is a position that we will fully flesh out throughout the rest of the book.

Talking about ancestry

In addition to kinship, nationalism, and embodied difference as cultural dimensions of relatedness, it is important to consider how people talk about ancestry because it can help us grapple with some of the popular and political debates surrounding ancestry (like the one we described in the Introduction). Just like nationalisms have to be represented, people have to talk about relatedness. It is important to pay attention to how people talk about relatedness because when a person talks about her ancestors, she will also be telling us something about who she is and how she sees herself in the world (this is in line with our discussion about time relativity above). Note that the forms of relatedness we have outlined above—kinship, nationalism, and embodied difference—are often tied to social problems and political issues. So even when people are "just talking about their ancestry," some deeper political meaning can be, and often is, at play. Different social groups have different positions of power, so when people talk about their relatedness, they are tying themselves to social groups that are in different positions of power.

Christopher Ball (2018) provides some useful guidelines for thinking about how people talk about kin relations. Paying attention to ethnographic context is critically important. For example, people thinking about African genetic ancestry in relation to slavery and reparations in the United States and the Caribbean (Benn Torres 2018) are linguistically engaged in a way that is quite different from, say, someone discussing at a dinner party the findings of their northern European ancestry revealed from a testing kit they received as a Christmas gift. Indeed, when people talk about genetic ancestry they are often, to varying degrees, trying to affirm or deny what others might consider a "normal" system of relatedness (Ball 2018, 49–53). How "normal" a system of relatedness operates depends on how many people in a social group accept that system. If, for example, there is a notion that all citizens of a nation are equal brothers and sisters yet significant social groups experience inequality, then the groups that experience inequality might not only reject the ideal of equality but also the notion that they are brothers or sisters with all citizens. Therefore, listening to how citizens refer to each other as "brother" or "sister" can tell us something about what they think about ideal of equal citizenship.

In addition to saying something about what is normal, when people talk about their ancestors they are often painting a much more complex worldview about *how their ancestors have existed through time*—from the past and into the present (Ball 2018, 53–54). People tend to think about their families, nations, and races as having certain characteristics—for example, biologically related, phenotypically similar, spiritual, legal, moral, emotionally predisposed in certain ways, and so on. These characteristics do not just apply to contemporary groups of relatives in a social group. Instead, the nature of these characteristics is derived through time but also projected from the present to the past. As such, how people talk about their ancestors is a way to both say something about the characteristics of those ancestors and about how their social group is or ought to be in the present. And finally, how a group ought to be in the present is usually tied to other political ideas about how one's social group stands relative to other social groups—in a community, country, or the world. For example, when a people talk about "a nation of immigrants," they might be making a statement about both how past national ancestors were immigrants and how contemporary society should treat immigrants. Or people who describe their contemporary social group as "a valiant race" could be both commenting on acts of bravery by their ancestors and the need for the contemporary social group to have more pride in themselves.

Conclusion

In this chapter we began by briefly exploring the philosophical complexities of the past in relation to time and space. The limitations that humans have with regard to thinking through time and space mean that time, space, and the past often have less to do with an attempt to absolutely grasp the entirety of known dimensions and more to do with how humans exist in the present. We held off on seriously entertaining the question of whether the past is real or not. Then, we delineated how humans experience relatedness in the forms of kinship, nationalism, and embodied difference; and we noted how those sociocultural time–space complexities manifested themselves when people talk about their ancestors. It is at this point in our argument that we can return to the profound philosophical question we put aside earlier regarding the reality of time.

Speaking about ancestry entails representing social groups within spaces and through time. Those representations are always partial because, as we noted earlier, humans cannot truly account for every bit of experience in their recollection of the past. Even the most accurate recollection of past events is still missing a lot of information. However, that missing information may be important to other people who have similar claims to the past; so the reality of an accurate account might not objectively address the past in a way that is important to people who have a vested interest in a shared time period. The reality of the past, therefore, is always relative to other people—that is, the past is social and cultural. Whether the past is real or not—philosophically speaking—is no more significantly real than the reality of how the past manifests itself in a sociocultural context. Indeed, the sociocultural realm is the most relevant reality to human existence because it takes into account both our evolutionary nature as social animals and our cultural ways of making meaningful sense out of the world that we inhabit. The question of how real the past is becomes a moot point once we realize that human's investment in the past is fundamentally sociocultural. Therefore, *genetic ancestry cannot represent a more accurate grasp on the past unless we consider its existence in a sociocultural context.* To put it differently, although genetic ancestry is based on science, the past is sociocultural, so even a scientific take on the past is relevant to human reality when considered in relation to a myriad of social and cultural conditions. Let us take the theoretically abstract claims in this paragraph and see them through the example at the beginning of this chapter.

Delroy claimed, "You will see that I am a son of Joseph Chatoyer!" This claim is more complex than a simple reaction to a request for his participating in a genetic ancestry research project. Delroy's claim has to be contextualized within several social groups to where he situates his relatives. He is from Rose Bank, a small community of families that are part of a broader set of communities tied to Garifuna identity. That Garifuna identity is not only specific to people who call themselves Garifuna, but it is also tied to a national sense of identity in St. Vincent. That sense of identity, which involves pride in a fighting spirit, is related to a broader set of colonial relationship with Great Britain, the former colonizer of St. Vincent. British forms of colonialism, in turn, are also intricately tied to ideas about embodied difference, specifically in the form of racial ideologies about indigeneity and blackness. Claiming to be the "son of Joseph Chatoyer," therefore, is a concrete example of how talk about relatedness is bundled with ideas about a historical past, a political present, and interconnected social groups.

At this point, we might ask: how might genetic ancestry figure into Delroy's claim? The answer is that genetic ancestry may not impact Delroy's sense of relatedness at all, it may impact this just a little, or it might change his whole worldview. If the genetic ancestry test happens to have some bearing on an aspect of a person's sense of identity that is very important to them, then it can be very impactful. However, if a person's sense of identity is not tied to conceptions of the past that are relevant to genetic ancestry (or if they just do not care about what genetic ancestry says), then there will be little or no impact. Moreover, because people's sense of relatedness spans social fields that have nothing to do with genetic relationships, we should never assume that genetic ancestry plays a significant social force. Of course, assuming that genetic ancestry has no social impact is also erroneous. The point is that we should carefully investigate the contexts in which genetic ancestry exists. Only then can we both comment on how we should think about genetic ancestry and resolve some contemporary debates regarding the role of genetic ancestry in contemporary societies.

References

Anderson, Benedict. 2006. *Imagined Communities: Reflections on the Origin and Spread of Nationalism*. Rev. ed. London: Verso.

Anderson, Edward. 2017. *The Problem of Time: Quantum Mechanics versus General Relativity*. 1st ed. New York: Springer.

Ball, C. 2018. "Language of Kin Relations and Relationlessness." *Annual Review of Anthropology* 47: 47–60. doi:10.1146/annurev-anthro-102317-050120.

Basso, Keith H. 1996. *Wisdom Sits in Places: Landscape and Language among the Western Apache*. Albuquerque: University of New Mexico Press.

Benn Torres, Jada. 2018. "'Reparational' Genetics: Genomic Data and the Case for Reparations in the Caribbean." *Genealogy* 2 (1): 7.

Benn Torres, Jada, and Gabriel A. Torres Colón. 2016. "Racial Experience as an Alternative Operationalization of Race." *Human Biology* 87 (4): 306–12.

Bouton, Christophe, and Philippe Huneman, eds. 2017. *Time of Nature and the Nature of Time: Philosophical Perspectives of Time in Natural Sciences.* New York: Springer.

Callender, Craig. 2011a. "Introduction." In *The Oxford Handbook of Philosophy of Time,* edited by Craig Callender, 1–10. Oxford: Oxford University Press. doi:10.1093/oxfordhb/9780199298204.001.0001.

———, ed. 2011b. *The Oxford Handbook of Philosophy of Time.* Oxford: Oxford University Press. doi:10.1093/oxfordhb/9780199298204.001.0001.

Carsten, Janet. 2003. *After Kinship.* Cambridge: Cambridge University Press. http://ebookcentral.proquest.com/lib/vand/detail.action?docID=259857.

Corrigan, Philip, and Derek Sayer. 1991. *The Great Arch: English State Formation as Cultural Revolution : [With Preface, Postscript and Bibliographical Supplement].* New York: Blackwell.

Egdall, Ira Mark. 2014. *Einstein Relatively Simple: Our Universe Revealed in Everyday Language.* Hackensack, NJ: World Scientific.

Gell, Alfred. 1992. *The Anthropology of Time: Cultural Constructions of Temporal Maps and Images.* Explorations in Anthropology. Oxford: Berg.

Gettler, Lee T., and James J. McKenna. 2011. "Evolutionary Perspectives on Mother–Infant Sleep Proximity and Breastfeeding in a Laboratory Setting." *American Journal of Physical Anthropology* 144 (3): 454–62. doi:10.1002/ajpa.21426.

Gettler, Lee T., Thomas W. McDade, Alan B. Feranil, and Christopher W. Kuzawa. 2011. "Longitudinal Evidence That Fatherhood Decreases Testosterone in Human Males." *Proceedings of the National Academy of Sciences* 108 (39): 16194–6199. doi:10.1073/pnas.1105403108.

Hill Collins, Patricia. 2004. *Black Sexual Politics: African Americans, Gender, and the New Racism.* New York: Routledge.

James, William. 2007. *The Principles of Psychology.* New York: Cosimo, Inc.

Kelly, John Dunham. 2001. *Represented Communities: Fiji and World Decolonization.* Chicago, IL: University of Chicago Press.

Lende, Daniel H., and Greg Downey. 2012. *The Encultured Brain: An Introduction to Neuroanthropology.* Cambridge: MIT Press.

Lucy, J. A. 1996. *Grammatical Categories and Cognition: A Case Study of the Linguistic Relativity Hypothesis.* Cambridge: Cambridge University Press.

Nagel, Joane. 2003. *Race, Ethnicity, and Sexuality: Intimate Intersections, Forbidden Frontiers.* New York: Oxford University Press.

Parkin, Robert, and Linda Stone, eds. 2004. *Kinship and Family: An Anthropological Reader.* Blackwell Anthologies in Social and Cultural Anthropology 4. Malden, MA: Blackwell Pub.

Peletz, Michael G. 1995. "Kinship Studies in Late Twentieth-Century Anthropology." *Annual Review of Anthropology* 24 (1): 343–72. doi:10.1146/annurev.an.24.100195.002015.

Rodseth, Lars. 1998. "Distributive Models of Culture: A Sapirian Alternative to Essentialism." *American Anthropologist*, New Series 100 (1): 55–69.

Sahlins, Marshall. 2000. *Culture in Practice: Selected Essays*. New York: Zone Books.

———. 2013. *What Kinship Is-And Is Not*. Chicago, IL: University of Chicago Press. http://ebookcentral.proquest.com/lib/vand/detail. action?docID=1094688.

Schneider, David Murray. 1980. *American Kinship: A Cultural Account*. 2nd ed. Chicago, IL: University of Chicago Press.

Torres Colón, Gabriel A. 2018. "Racial Experience as Bioculturally Embodied Difference and Political Possibilities for Resisting Racism." *The Pluralist* 13 (1): 131–42.

3 Genetic perspectives on the past

Over the course of our fieldwork experiences, we learned that people's motivations for participating in an ancestry study vary, as do the ways they engage with their ancestry results on our return visits. When we meet participants, what is clear is that they all have ideas about the past and relatedness that have been shaped by their understandings of the world, individual experiences, and their socialization within their communities. Participants' ideas about genetic technologies are also relevant in thinking about how they, or any ancestry test-taker, values genetic ancestry. In this chapter, we discuss how genetic ancestry technology, inclusive of test results, can affect how people think about their pasts and how people's existing notions about the past affect the way they give meaning to genetic ancestry technology. Drawing from the work of other scholars and our own fieldwork, the point we make, a point that should resonate for the remainder of this book, is that *there is no natural relationship between genetic ancestry technology and how people interpret genetic information*. Rather, the sociocultural context in which people encounter genetic ancestry provides a prism for what genetic ancestry means and ultimately what value it carries. This point, as we shall see later in the chapter, is often lost when academics launch critiques about genetic ancestry technologies—even when such critiques are intellectually sound and academically necessary.

Giving meaning to genetic ancestry technology

In thinking about how people interpret genetic ancestry, it is important to consider the various sociocultural dimensions of relatedness that we discussed in Chapter 2: kinship, nationalism, and embodied difference. These dimensions help to articulate the context surrounding genetic ancestry technology and can provide some insight into the role or value that this particular technology holds for interested parties. Anthropologists and sociologists have employed several methods, varying from ethnography to surveys, to document the varied meanings and contexts surrounding genetic ancestry. Here we rely on published research in addition to our own fieldwork to make a cross-cultural assessment of how genetic ancestry is both influential and influences people's ideas about the past. As we shall see, cross-cultural insights are useful for understanding how, under different conditions, racial inequity, political ideologies, and institutional privilege of science can affect people's experience with genetic ancestry research and direct-to-consumer testing.

We begin our cross-cultural exploration of the multitude of ways that genetic ancestry is experienced using several ethnographic case studies from colleagues working throughout the Americas. Sarah Abel (2018), highlights the emergence of different sentiments about genetic ancestry within African descendant communities from the United States, the French Antilles, and in Brazil. Through the use of complementary work by other researchers in addition to ethnographic methods, Abel finds that the acceptance or rejection of genetic ancestry technology as a way to examine the past is, "… shaped by various overarching sources of influence notably national narratives that are bound up with ideologies of race, anti-racism, and concept of citizenship; discourses of scientists and experts regarding genetic and genealogical knowledge; and projects of activists, aimed at fostering community healing, race consciousness and political protest." (p. 17). Using cross-cultural comparisons, she observes that when looking at how communities value genetic ancestry as a tool to meet their personal or political agendas regarding reconciling the past, communities reach different conclusions because of different sociocultural, political, and historical factors.

In her consideration of North American African Americans, Abel discusses how some African Americans are very interested in using genetic ancestry as a means to identify African origins and cultivate a sense of belonging beyond the United States. As Abel reports it, for some African Americans, the need to locate African origins emanates from a general American interest in genealogy as illustrated by the ubiquitous family tree projects given to school children, the emergence of genealogy resources as the Internet became more commonplace, the publication of Alex Haley's *Roots,* as well as other related factors (Abel 2018). For many African Americans, tracing family history using conventional genealogical methods is complicated due to the limited records kept about enslaved peoples. However, the advent of commercially available genetic ancestry testing brought about a new way of addressing questions of origin (Nelson 2008b). In Abel's study, some of her participants reported seeking out genetic ancestry tests with the express purpose of learning about family origins prior to the Transatlantic slave trade. Abel also observed a community of African American genealogists that placed a greater utility on genetic testing as a means to identify close relatives. For these genealogists, identifying more recent familiar histories was more important than finding connections to distant homelands. These genealogists wanted to engage with genetic technologies in so far as it provided answers that conventional genealogical methods could not. Unlike the African American participants that wanted to learn more about their African origins, these genealogists were more focused on using genetic technologies to build and supplement family history. For these genealogists, there reportedly was not a strong need to know more about their ancestry beyond the regions they and recent past generations called home. The difference between how the groups of African Americans in Abel's study accept or reject genetic ancestry highlights the point that differing political and personal agendas influence how genetic ancestry technologies are valued.

The French Antilles was the second focus of Abel's cross-cultural comparison of genetic ancestry technologies. She used primary and secondary sources to consider what genetic ancestry testing means among some French Antillean communities (communities from the Caribbean islands of Guadeloupe and Martinique). Abel found that the sentiment about genetic ancestry among these individuals contrasted with that observed in the United States. Among the French Antillean people considered in Able's study, genetic ancestry was not prioritized for its ability to be informative about African or more recent ancestry. Instead, methods or techniques that would help to make connections to the emergence of French Antillean communities were prioritized due to participants' desire to connect to the Americas, specifically the colonial history of the Americas. Given this interest, people were most concerned with identifying or connecting to but not beyond their enslaved ancestors in the Antilles. This effectively meant that genetic ancestry technologies that focus on continental ancestry from many generations ago were not particularly useful because this type of genetic testing does not provide the type of information within the timeframe that they sought. The last part of Abel's cross-cultural comparisons regarding the use of genetic ancestry technologies considered Brazil. According to Abel's ethnographic work, the sentiments of activists and other interested people assessing this technology in Brazil are complex, at times conflicting, and differing from that observed among peers in the United States and in the French Antilles.

Brazil has a national narrative that promotes a history marked by racial mixing, also referred to as *mestiçagem* (Mitchell 2017). In addition, through policy and sociocultural practices, preference for lighter skin and physical attributes characteristic of European peoples, referred to as "social whitening," have been idealized within Brazilian society (dos Santos 2002; Telles 2004; Mitchell 2017). As explained by Abel and others, this particular history has resulted in a suppression of blackness in Brazil, meaning that the history and influence of African peoples in Brazil has been conventionally under-acknowledged and obscured (dos Santos 2002; Telles 2004; Caldwell 2007). Though due to more recent legislation, for example Law 10.639, passed in 2003, which made teaching Afro-Brazilian history obligatory in public and private schools, this practice has begun to change (Gomes 2015; Guimarães 2015; dos Santos and Soeterik 2016). Nevertheless, the combination of the ideas of mestiçagem and "social whitening" has created, for some self-identified Afro-Brazilians, situations in which information about their African origins is limited due to social convention of prioritizing non-African origins. Additionally, with the backdrop of mestiçagem, some self-identified Afro-Brazilian communities struggle to be recognized as Afro-Brazilian yet simultaneously suffer the social and economic consequences of being marginalized for their blackness. In this case, genetic ancestry technologies turn out to be both a solution and a problem. Because genetic ancestry testing can be informative about ancestral origins, there is the potential for test-takers to learn about their family histories that have otherwise been obscured for lack of connection to Europe. Here genetic ancestry joins the tool kit for building resilience and becomes a tool for empowerment and resistance against the under-acknowledgment of blackness in Brazil (Hordge-Freeman 2013). On the other hand, these same genetic ancestry tests may well reveal appreciable amounts of non-African admixture and thus can potentially undermine claims to one particular portion of their ancestry, African ancestry, as primary in their social identities, that is as Afro-Brazilians. According to this logic and within the context of mestiçagem, the prioritization of African ancestry among self-identified Afro-Brazilians that have some proportion of non-African ancestry is problematic because identifying in this way ignores and perhaps devalues the genetic contributions from non-African ancestors (dos Santos 2006). The internal conflict regarding the politics of self-identification is not lost among Brazilians as a whole, rather as recorded by a variety of researchers, the debate of having ancestors that were both the victims and perpetrators of a racist past is ongoing and impacts the policies affecting the lives of Brazilians today (Harris et al. 1993; Htun 2004; Paschel 2016). Beyond identity politics, the mestiçagem narrative and ancestry test results therefore can complicate political agendas that work toward creating equity for Afro-Brazilians. For example, the use of affirmative action policies designed to assist in creating additional equitable opportunities between marginalized and non-marginalized peoples (Htun 2004; Mitchell 2017) is complicated because genetic technologies do not and cannot provide a clear way to determine who is black, who is not black, and therefore who is eligible for such policies. In Abel's study about the sentiments of genetic ancestry technologies in Brazil, she finds that potential test-takers are extremely wary of the technology despite the potential to address gaps in knowledge about connections to Africa. The hesitation to use the technology, it seems, emanates from the potential for harm to social justice agendas. In this case, genetic ancestry technologies are at the crux of two competing agendas for Afro-Brazilians; genetic ancestry testing can help to recover a past that has been denied to them and simultaneously used to deny futures filled with equitable opportunities. Ultimately, for many Afro-Brazilians, it seems that genetic ancestry technologies may remain as contested as the issues of race and identity politics.

The scenarios described in Abel's work illustrates some of the many ways that genetic ancestry technologies are referenced for reconciling the past and making sense of the present. These examples, however, do not encompass all of the sentiments about genetic ancestry technologies that we have encountered in our fieldwork. This final example comes from our fieldwork among Accompong Town Maroons in rural Jamaica. In this case, for one study participant, genetic ancestry was not regarded as particularly important or unimportant in understanding the past, rather it was not regarded at all.

Accompong Town is located in the southwestern part of Jamaica, approximately 31.34 km (19.5 miles) south of Montego Bay. Many of the people currently residing in Accompong Town descend from Africans that were known as Maroons. The name "Maroons" was used to describe individuals that defied colonial systems, rejected enslavement, and through war with the British won a peace treaty in 1739, establishing semi-independent communities within the Jamaican hinterland (Kopytoff 1978). Today, Maroon descendant communities are found primarily in two rural centers in Jamaica. The Windward Maroon settlements are found in the eastern part of Jamaica near the Blue Mountains with settlements in St. Mary's Parish and St. Ann's Parish. The Leeward Maroon settlements are located in the western section of the island including the parishes of St. James, St. Elizabeth, and Trelawney (Agorsah 1994; Brandon 2004). The Accompong Town Maroons belong to the Leeward settlements. While scholars and other historians of Jamaican history contend that Maroons can trace their origins to only African peoples (Campbell 1988), the Accompong Town Maroons have oral history that indicates partial ancestry from Jamaica's Indigenous population (Carey 1997). According to some Maroons, the native peoples of Jamaica were the first Maroons because they took to the hinterlands to escape persecution by European colonists and were later followed by African peoples (Kopytoff 1976).

In the summer of 2011, I (Jada) initiated a study to use genetic ancestry to examine the origins of the Accompong Town Maroons while Gabriel investigated Maroon youth identity. The genetic ancestry study revealed that Accompong Town Maroons indeed can trace most of their ancestry to West African peoples; however, they also can trace their ancestry to the Indigenous people of the Americas (Madrilejo, Lombard, and Benn Torres 2015; Fuller and Benn Torres 2018). After the analysis was done, we made a return visit to the community to update participants about the findings of the study, to return individualized results of their ancestry tests, and to learn how they understood and processed the study findings. In addition, we wanted to experience the Accompong Town Maroon Independence Day festival held annually on January 6 (Djedje 1998). I was particularly eager to speak to the people whose mitochondrial haplogroup revealed Indigenous American ancestry. In particular, I was hoping to learn more about these particular families' histories and what these participants thought about having maternal Indigenous American ancestry. As fate would have it, one of these participants left Accompong in search of better job opportunities and consequently was lost to follow-up. However, I was fortunate to find the other participant, Selwyn, three days into our week-long stay in Accompong Town.

On the day I finally caught up with Selwyn, he was standing near a car in the front yard of a home with several other Accompong residents. As I approached him, I handed over the envelope containing the study findings, the personalized ancestry test results, and very excitedly began to recap the study findings. I could see that Selwyn and the others were on their way to some other destination, so I hurriedly relayed the findings. Selwyn listened, nodded, and politely smiled at me. I began to explain the test results, building up my explanation of genetics and ancestry all the while anticipating some sort of surprise or shock once I revealed everything, especially the part about Selwyn's Indigenous American ancestry. To my surprise and shock, Selwyn continued to smile politely and did nothing more than acknowledge what I said and thanked me for the information. Grasping the envelope, Selwyn climbed into the car with the other people. At that moment, I suddenly understood that the follow-up interview I had hoped for was not going to happen. As a last-ditch effort, I asked Selwyn to meet sometime later that week. Selwyn continued to nod and smile but made no effort to agree on a next meeting time or place. The car drove away as did my chance to learn more about Selwyn's family history. Needless to say, I did not see Selwyn again. I was left standing empty-handed in the yard wondering if I was misunderstood and if Selwyn realized just how important the test findings were because they aligned with Maroon oral history. Mostly, however, I wondered how Selwyn was not as excited as I was about the genetic ancestry results. I will likely never know Selwyn's motivations for joining my study in the first place, his sentiments about the specific ancestry results, or his ideas about genetic ancestry testing in general. Based on Selwyn's response, it appears that genetic ancestry is not necessarily central or even at the periphery of his sense of the past. As I was told by another Maroon community member who declined to participate in my project during the recruitment phase, "I don't need that [DNA collection swab] to know about Maroons, I know my history."

The examples discussed in Abel's work among peoples from the United States, the French Antilles, and Brazil as well as my work in Jamaica reiterate the point that there is no one specific way that people interact and interpret genetic ancestry tests and the past. Rather, the ways that people reconcile the past using genetic technologies is influenced by a number of competing factors including historical factors, sociocultural contexts, and contemporary political agendas. Several other researchers have come to similar conclusions, noting the variable but limited impact that genetic ancestry testing appears to have on how test-takers interpret the results (Nelson 2008a, 2016; Schramm, Skinner, and Rottenburg 2012; Lee 2013; Scully, Brown, and King 2016; Shim, Alam, and Aouizerat 2018; Benn Torres 2018). The examples discussed here as well as in the cited literature illustrate that genetic ancestry technologies are void of meaning until it is given some sort of meaning by people interested in using or setting aside this technology.

Noting that there is no natural relationship between genetic ancestry technology and how test-takers interpret genetic information, we must still acknowledge that the weight of science behind genetic ancestry is hard to ignore. Genetic ancestry testing is designed and marketed specifically as a scientific technology that may be used to assess some quality of test-takers (e.g., ancestry or relatedness to other test-takers; Bolnick et al. 2007; Royal et al. 2010). As such, academics have leveled heavy critiques against genetic ancestry technologies and the ways that many geneticists discuss the intersections between genetic ancestry and society, namely what genetic ancestry means with regard to race (Fullwiley 2008; Duster 2009, 2015; Reardon 2011; Reardon and TallBear 2012; TallBear 2013). To a large degree, these critiques are justifiable and arguably should encourage geneticists to be more critical of their implicit assertions linking genetics to race. However, these critiques also tend to discount the ways that genetic ancestry may be leveraged in ways that disrupt biological race concepts and how genetic ancestry can work as potential sources of community empowerment.

Clines, substructure, and other limitations: scholarly critiques of genetic ancestry

Within and among the biological and social sciences, there is still considerable controversy about how best to describe human genetic variation as well as describing how this variation, in regards to genetic ancestry, relates to experience including race. Anthropologists agree that contemporary human genetic variation is the result of our evolutionary history that is marked by a common origin followed by serial founder effects, intermittent isolation due to geography, and non-random mate choice as a result of sociocultural factors (Jobling, Hurles, and Tyler-Smith 2013; Stoneking 2017; Relethford and Bolnick 2018). As discussed in Chapter 1, the outcome is a specific pattern of variation where, on the one hand, human variation is clinally distributed across geographic space. On the other hand, human variation can be queried in such a way that substructure, or relatively homogenous sub-clusters of human groups, becomes a defining characteristic when considering geographically distant regions (Jobling, Hurles, and Tyler-Smith 2013). These sub-divisions are a direct function of the relationship between geography and genetic distance.

Genetic ancestry testing relies on substructure to assign test-taker's ancestry to broad continental regions. However, characterizing human genetic variation as primarily substructured has led some social scientific scholars to assert that genetic ancestry testing aligns with contemporary iterations of scientific racism (Carter 2007; El-Haj 2007; Fullwiley 2008). According to these critiques, portraying human genetic variation as primarily substructured supports the idea that humans can be systematically distributed in discrete biological clusters, and these biological clusters tend to align with colloquial understandings of racial groups. These critiques were largely based on early genetic work that looked at the distribution of human variation across global populations. One study in particular, Rosenberg and colleagues (2002), examined patterns of genetic variation in 52 populations from around the world. The findings of this study suggested that the vast majority, between 93% and 95%, of genetic variation could be found within populations while the remaining 3–5% could be attributed to differences between groups. This means that there is more variation within groups than between groups and attests to the overall genetic similarity of all humans. However, when looking at the broad distribution of genetic variation across these global groups, they found the most statistical support for five geographic clusters corresponding to the following continental regions: Africa, the Americas, East Asia, Europe, and Oceania. Though not promoted by the authors as such, the connection between the identified bio-geographic clusters and conventional ideas about biological races was quickly noted in the popular press (Wade 2002, 2003). Based upon these findings, it un-derstandable how someone might think of the observed substructure as reflecting real biological differences between geographic groups, which in turn equate to biological differences between racial groups. Social scientists critiqued this research noting that when genetic variation is queried to highlight substructure vis-à-vis genetic ancestry testing, the concept of biological races is reified, meaning the idea that there are biologically distinct races becomes more concrete, more real. The reification of biological races diametrically opposes the anti-racist traditions that mark contemporary anthropology. Contemporary anthropological takes on human variation state that while human variation exists and is important to consider, humans cannot reasonably be classified into biologically defined racial groups (AAPA Committee on Diversity 2019).

While it is entirely feasible that the ways in which genetic ancestry is presented and marketed make biological concepts of race more concrete, how much people reify biological race through genetic ancestry is an empirical question in need of research. One can ask, against the backdrop of a test-taker's ethnic or racial socialization, do genetic ancestry tests actually reify race? Several researchers have conducted survey studies that examine test-takers' ideas about race and self-identification both in the United States and in Brazil (Santos et al. 2009; Jonassaint et al. 2010; Wagner 2010; Wagner and Weiss 2011; Baptista et al. 2016; Shim, Alam, and Aouizerat 2018). The emerging conclusion from these studies is that genetic ancestry tests have little to no effect on how people think about race or how they self-identify.

Rather, test-takers recognize that other factors, such as skin color or prevailing sociocultural models of race, tend to be more formative in their understanding and experience of race than genetic ancestry results. Santos and colleagues (2009) succinctly state this finding in their study of the impact of genetic ancestry testing on ideas of race and affirmative action policies among Brazilian youth:

> The dominant view was that the genetic results, whatever they were, would not have much influence. "It is important to know about them, but, beyond this, this knowledge is not worth anything at all"; "this business of tests is a nice curiosity, and so what? … I'll bet that when I apply they won't take the test results into account…. In spite of that high percentage of European ancestry I won't cease to be 'black'; never!" (Santos et al. 2009, 796).

We include these studies not to contradict the concern that many social scientists have that scientific racism is alive and well. On the contrary, we note that there is also very good evidence that genetic ancestry testing is leveraged by some to uphold biological notions of race. For example, Nicholas Wade's publication *A Troublesome Inheritance: Genes, Race and Human History* (2014) was condemned as misusing research by geneticists and other scholars (Caspari 2014; Coop et al. 2014; Fuentes 2014) as well as the use of genetic ancestry testing by white supremacist groups (Boodman 2017; Zhang 2017; Harmon 2018; Panofsky and Donovan 2019). The broader point we intend to convey is that there is no natural relation between genetic ancestry and interpretation, and furthermore that genetic ancestry testing is a tool. Like any tool, it can be used to build or to break. For example, despite associations with scientific racism, genetic ancestry can also be used to undermine the same ideas it has been accused of perpetuating. In these instances, genetic ancestry serves to disrupt biological notions of race, dismantle social constructions of racial and national distinctiveness, and instead highlight the biological unity of the human species. One way this has been done has been to draw attention to the nonconcordance between genetic ancestry results and self-identification. In these types of exercises, the social dimensions of racial constructs become that much more apparent. Published online in June 2016, a series of videos called *The DNA Journey* was produced by a travel app search engine company, momondo. The videos went viral and, according to the momondo website, they were viewed over 28 million times on Facebook alone. In these videos, prior to taking an ancestry test, participants were asked about their genealogies. Some of the participants inserted very nationalistic views into their description of their family history. The videos go on to show the participants' reactions after the revelation of their genetic ancestry results. The surprise in their reactions as well as the subsequent reconsideration of their ideas about human difference works toward the culminating point of the video series: "An open world begins with an open mind" "Let's open our world." The videos work to illustrate the unity of the human species and in doing so challenges viewers to recognize that the emphasis made on differences between people are indeed constructed and not supported with biological data. A 2015 study conducted by scientists associated with research institutions and also with direct-to-consumer (DTC) ancestry testing companies came up with similar findings in their examination of self-identified race and genetic ancestry among 23 &Me customers (Bryc et al. 2015). In this study they noted the complexities surrounding how people self-identified in relation to their genetic ancestry proportions. Bryc and colleagues concluded:

> Perhaps most importantly, however, our results reveal the impact of centuries of admixture in the US, thereby undermining the use of cultural labels that group individuals into discrete nonoverlapping bins in biomedical contexts "which cannot be adequately represented by arbitrary 'race/color' categories."
>
> (Bryc et al. 2015, 50)

Academics, ourselves included, have also employed genetic ancestry to displace hegemonic narratives regarding marginalized communities. For example, for some members of the Indigenous Caribbean communities we work with as well as other Caribbean peoples, genetic ancestry estimates illustrating Indigenous American ancestry works toward disproving narratives of indigenous extinction and highlight the roles that people from Africa, Asia, Europe, and Indigenous America have had in shaping the contemporary region (Benn Torres 2014, 2018). As described previously in this chapter, genetic ancestry can provide an additional perspective to thinking about the past and how the past influences the present. We argue that in cases similar to those among Indigenous Caribbean or among Afro-Brazilians, genetic ancestry does not have to reify race but instead can work to dismantle ideas that people already had about race and difference.

As we move toward the end of this discussion on genetic ancestry and notions of the past and present, we want to emphasize a point which has to do more with the prioritization of certain ways of knowing or epistemologies as opposed to ancestry tests themselves. Genetic ancestry tests can only ascertain information about ancestry based on biological notions of relatedness. Human groups have exhibited a multitude of ways to create, build, and maintain relationships with one another that transcend biology (Marks 2013). Genetic relationships are only one of many ways that people relate to each other. However, people form and value spiritual connections, adoptions, shared worldviews, and other means of creating bonds and building relationships— all of which are beyond the purview of genetic ancestry tests. Denying, disregarding, or devaluing these types of relationships in place of biological notions of relatedness does a disservice to the diverse ways in which humans experience being human. This point has been especially poignant in genetic ancestry work among marginalized communities where political agendas and issues of sovereignty potentially corroborate or sit at odds with genetic ancestry results and interpretations (Reardon and TallBear 2012; TallBear 2013; Bolnick et al. 2016; Nelson 2016; Benn Torres 2018). Recent examples of this may be found among the native peoples (Taíno) of the Greater Antilles as well as Afro-Indigenous peoples of North America. In the case of the Greater Antilles, for some self-identified Taíno, genetic ancestry helps to contend narratives of their supposed extinction (Feliciano-Santos 2011; Benn Torres 2014; Bukhari et al. 2017; Fuller and Benn Torres 2018). Among some self-identified Afro-Indigenous communities in North America, genetic ancestry complicates their ideas about the relationships between their ancestors and consequently their relationship with contemporary sovereign indigenous communities (Johnston 2003; Cooper 2017). This critique regarding genetic ancestry as a way of knowing about the past is worthy of discussion because of the weight of science and subsequent prioritization that genetic ancestry receives precisely because it is perceived as scientific. However, the prioritization of genetic ancestry must also be weighed against histories of forced migrations, genocides, enslavement, and systematic discrimination. These factors have shaped the understood and lived relationships within and between communities and these relationships may not be apparent using genetic ancestry tests. Nonetheless, it is important to note that despite the capabilities of ancestry tests, non-biologically based relationships are also valid and valued relationships.

Competing ideologies

We end this chapter with a discussion of the epistemological limitations of genetic ancestry not to discredit ancestry testing or more broadly science as a way of knowing, but to draw attention to the fact that despite the objectivity built into science as a way to systematically learn about the world, science is still predisposed to the subjectivities of those that practice it (Marks 2009). Ways of organizing and understanding the world shape how science and scientific questions are created and queried. Marvin Harris, a champion of scientific anthropology, often acknowledged that science is an ideology, just like any other human ideology, but one that happens to have a special relationship with reality external to human perception. His view is both anthropological and scientific. It is anthropological because it views science through a cultural lens. Science is comprised of culturally held ideas by groups of peoples in specific times and space. Calling science an ideology simply means that it is comprised of collectively held ideas and behaviors that exist within webs of meaning and signification. Claiming that science is an ideology that has a special relationship with reality external to human perception also takes a firm realist stance on the ability of science to comment on the world around us in ways that are more likely to reach consensus across cultures.

We need not enter a debate, much less resolve a debate, about the nature of truth, science, and culture in order to make the point that we want to get across in this chapter. Mainly, that "the past" and "our ancestors" are both ideas that are not particular to either science or any other cultural ideologies. Of course, we do not want to argue that these ideas about the past are freely and equally floating around in society for individuals to subscribe to one or another. Ideas about the past are often part of broader ideas about "who we are"; in turn, "who we are" is often part of a political tool for organizing and controlling economic relationships in a society. For example, the idea of a "nation" is tied to popular notions of "who we are," and people in positions of power often occupy a position of privilege in articulating the history of the people who comprise a nation. Scientists and academics also occupy a position of privilege when articulating ideas about the past through their work. We think that the best academic work about the past takes into account the knowledge held by the people whose past academics write about. In the next chapter, we discuss in more depth race and genetics, a topic in which scientists and other academics have wielded their privilege and shaped discourse.

References

AAPA Committee on Diversity. 2019. "AAPA Statement on Race & Racism." http://physanth.org/about/position-statements/aapa-statement-race-and-racism-2019/.

Abel, Sarah. 2018. "Of African Descent? Blackness and the Concept of Origins in Cultural Perspectives." *Genealogy* 2 (1): 11. doi:/10.3390/genealogy 2010011.

Agorsah, E. K. 1994. *Maroon Heritage: Archaeological, Ethnographic, and Historical Perspectives.* Kingston: University of the West Indies Press.

Baptista, Natalie M., Kurt D. Christensen, Deanna Alexis Carere, Simon A. Broadley, J. Scott Roberts, and Robert C. Green. 2016. "Adopting Genetics: Motivations and Outcomes of Personal Genomic Testing in Adult Adoptees." *Genetics in Medicine* 18 (9): 924–32. doi:10.1038/gim.2015.192.

Benn Torres, Jada. 2014. "Prospecting the Past: Genetic Perspectives on the Extinction and Survival of Indigenous Peoples of the Caribbean." *New Genetics and Society* 33: 21–41.

———. 2018. "'Reparational' Genetics: Genomic Data and the Case for Reparations in the Caribbean." *Genealogy* 2 (1): 7. doi:10.3390/genealogy2010007.

Bolnick, Deborah Ann, Jennifer A. Raff, Lauren C. Springs, Austin W. Reynolds, and Aida T. Miró-Herrans. 2016. "Native American Genomics and Population Histories." *Annual Review of Anthropology* 45 (1): 319–40. doi:10.1146/annurev-anthro-102215-100036.

Boodman, Eric. 2017. "White Nationalists Are Flocking to Genetic Ancestry Tests--With Surprising Results." *Scientific American.* August 16, 2017. https://www.scientificamerican.com/article/white-nationalists-are-flocking-to-genetic-ancestry-tests-with-surprising-results/.

Brandon, George. 2004. "Jamaican Maroons." In *Encyclopedia of Medical Anthropology: Health and Illness in the World's Cultures Volume I: Topics Volume II: Cultures,* edited by Carol R. Ember and Melvin Ember, 754–65. Boston, MA: Springer US. doi:10.1007/0-387-29905-X_77.

Bryc, K., E. Y. Durand, J. M. Macpherson, D. Reich, and J. L. Mountain. 2015. "The Genetic Ancestry of African Americans, Latinos, and European Americans across the United States." *American Journal of Human Genetics* 96: 37–53. doi:10.1016/j.ajhg.2014.11.010.

Bukhari, Areej, Javier Rodriguez Luis, Miguel A. Alfonso-Sanchez, Ralph Garcia-Bertrand, and Rene J. Herrera. 2017. "Taino and African Maternal Heritage in the Greater Antilles." *Gene* 637 (December): 33–40. doi:10.1016/j.gene.2017.09.004.

Campbell, Mavis Christine. 1988. *The Maroons of Jamaica, 1655–1796: A History of Resistance, Collaboration & Betrayal.* South Hadley, MA: Bergin & Garvey.

Carey, Bev. 1997. *The Maroon Story: The Authentic and Original History of the Maroons in the History of Jamaica, 1490–1880* (A Maroon and Jamaica Heritage Series). St. Andrew, Jamaica: Agouti Press.

Carter, Robert. 2007. "Genes, Genomes and Genealogies: The Return of Scientific Racism?" *Ethnic and Racial Studies* 30 (July): 546–56. doi:10.1080/01419870701355983.

Caspari, Rachel. 2014. "A Troublesome Inheritance: Genes, Race and Human History by Nicholas Wade." *American Anthropologist* 116 (4): 896–97. doi:10.1111/aman.12162_32.

Coop, Graham, Micheal B. Eisen, Rasmus Nielsen, Molly Przeworski, and Noah Rosenburg. 2014. "Letters: 'A Troublesome Inheritance.'" *The New York Times*, August 8, 2014, sec. Books. https://www.nytimes.com/2014/08/10/books/review/letters-a-troublesome-inheritance.html.

Cooper, Kenneth J. 2017. "Perspective | I'm a Descendant of the Cherokee Nation's Black Slaves. Tribal Citizenship Is Our Birthright." *Washington Post*, September 15, 2017, Post Nation edition. https://www.washingtonpost.com/news/post-nation/wp/2017/09/15/im-a-descendant-of-the-cherokee-nations-black-slaves-tribal-citizenship-is-our-birthright/.

Djedje, Jacqueline Cogdell. 1998. "Remembering Kojo: History, Music, and Gender in the January Sixth Celebration of the Jamaican Accompong Maroons." *Black Music Research Journal* 18 (1/2): 67–120.

dos Santos, Renato Emerson Nascimento, and Inti Maya Soeterik. 2016. "Scales of Political Action and Social Movements in Education: The Case of the Brazilian Black Movement and Law 10.639." *Globalisation, Societies and Education* 14 (1): 30–48. doi:10.1080/14767724.2015.1051000.

dos Santos, Sales Augusto. 2002. "Historical Roots of the 'Whitening' of Brazil." *Latin American Perspectives* 29 (1): 61–82. doi:10.1177/0094582X0202900104.

———. 2006. "Who Is Black in Brazil? A Timely or a False Question in Brazilian Race Relations in the Era of Affirmative Action?" *Latin American Perspectives* 33 (4): 30–48. doi:10.1177/0094582X06290122.

Duster, Troy. 2011. "Ancestry Testing and DNA : Uses, Limits, and Caveat Emptor." In *Race and the Genetic Revolution: Science, Myth, and Culture*, by Evelynn M. Hammonds, edited by Sheldon Krimsky and Kathleen Lewis Sloan, 99–115. New York: Columbia University Press.

———. 2015. "A Post-genomic Surprise. The Molecular Reinscription of Race in Science, Law and Medicine." *The British Journal of Sociology* 66: 1–27.

El-Haj, Nadia Abu. 2007. "The Genetic Reinscription of Race." *Annual Review of Anthropology* 36: 283–300. doi:10.1146/annurev.anthro.34.081804.120522.

Feliciano-Santos, Sherina. 2011. "An Inconceivable Indigeneity: The Historical, Cultural, and Interactional Dimensions of Puerto Rican Taino Activism." Ann Arbor: University of Michigan.

Fuentes, Agustín. 2014. "A Troublesome Inheritance: Nicholas Wade's Botched Interpretation of Human Genetics, History, and Evolution." *Human Biology* 86 (3): 215–19. doi:10.13110/humanbiology.86.3.0215.

Fuller, Harcourt, and Jada Benn Torres. 2018. "Investigating the 'Taíno' Ancestry of the Jamaican Maroons: A New Genetic (DNA), Historical, and Multidisciplinary Analysis and Case Study of the Accompong Town Maroons." *Canadian Journal of Latin American and Caribbean Studies/Revue Canadienne Des Études Latino-Américaines et Caraïbes* 43 (1): 47–78.

Fullwiley, Duana. 2008. "The Biologistical Construction of Race: 'Admixture' Technology and the New Genetic Medicine." *Social Studies of Science* 38: 695–735. doi:10.1177/0306312708090796.

Gomes, Nilma Lino. 2015. "Making the Teaching of Afro-Brazilian and African History and Culture Compulsory: Tensions and Contradictions for Anti-Racist Education in Brazil." *In Eurocentrism, Racism and Knowledge*, edited by Marta Araújo and Silvia Rodríguez Maeso, 192–208. New York, N.Y.: Palgrave Macmillan. link.springer.com/chapter/10.1057/9781137292896_11#citeas.

Guimarães, Selva. 2015. "The Teaching of Afro-Brazilian and Indigenous Culture and History in Brazilian Basic Education in the 21st Century." *Policy Futures in Education* 13 (8): 939–48. doi:10.1177/1478210315579980.

Harmon, Amy. 2018. "Why White Supremacists Are Chugging Milk (and Why Geneticists Are Alarmed)." *The New York Times*, October 19, 2018, sec. U.S. https://www.nytimes.com/2018/10/17/us/white-supremacists-science-dna.html.

Harris, Marvin, Josildeth Gomes Consorte, Joseph Lang, and Bryan Byrne. 1993. "Who Are the Whites?: Imposed Census Categories and the Racial Demography of Brazil." *Social Forces* 72: 451–62. heinonline:/HOL/P?h=hein.journals/josf72&i=468.

Hordge-Freeman, Elizabeth. 2013. "What's Love Got to Do with It?: Racial Features, Stigma and Socialization in Afro-Brazilian Families." *Ethnic and Racial Studies* 36 (10): 1507–23. doi:10.1080/01419870.2013.788200.

Htun, Mala. 2004. "From 'Racial Democracy' to Affirmative Action: Changing State Policy on Race in Brazil." *Latin American Research Review* 39 (1): 60–89. www.jstor.org/stable/1555383.

Jobling, Mark, Matthew Hurles, and Chris Tyler-Smith. 2013. *Human Evolutionary Genetics: Origins, Peoples & Disease*. New York: Garland Science.

Johnston, Josephine. 2003. "Resisting a Genetic Identity: The Black Seminoles and Genetic Tests of Ancestry." *The Journal of Law, Medicine & Ethics* 31: 262–71. doi:10.1111/j.1748-720X.2003.tb00087.x.

Jonassaint, Charles R., Eunice R. Santos, Crystal M. Glover, Perry W. Payne, Grace-Ann Fasaye, Nefertiti Oji-Njideka, Stanley Hooker, et al. 2010. "Regional Differences in Awareness and Attitudes Regarding Genetic Testing for Disease Risk and Ancestry." *Human Genetics* 128 (3): 249–60. doi:10.1007/s00439-010-0845-0.

Kopytoff, Barbara. 1976. "The Development of Jamaican Maroon Ethnicity." *Caribbean Quarterly: Essays on Slavery* 22 (2–3): 33–50. doi:10.1080/00086495.1976.11671900.

———1978. "The Early Political Development of Jamaican Maroon Societies." *The William and Mary Quarterly* 35 (2): 287–307. doi:10.2307/1921836.

Lee, Sandra Soo-Jin. 2013. "Race, Risk, and Recreation in Personal Genomics: The Limits of Play." *Medical Anthropology Quarterly* 27 (4): 550–69. doi:10.1111/maq.12059.

Madrilejo, Nicole, Holden Lombard, and Jada Benn Torres. 2015. "Origins of Marronage: Mitochondrial Lineages of Jamaica's Accompong Town Maroons." *American Journal of Human Biology* 27 (3): 432–437. https://doi.org/doi.org/10.1002/ajhb.22656.

Marks, Jonathan. 2009. *Why I Am Not a Scientist: Anthropology and Modern Knowledge*. Berkeley: University of California Press.

———. 2013. "The Nature/Culture of Genetic Facts*." *Annual Review of Anthropology* 42: 247–67. doi:10.1146/annurev-anthro-092412-155558.

Mitchell, Sean T. 2017. "Whitening and Racial Ambiguity: Racialization and Ethnoracial Citizenship in Contemporary Brazil." *African and Black Diaspora: An International Journal* 10 (2): 114–30. doi:10.1080/17528631.2016.1189693.

Nelson, Alondra. 2008a. "The Factness of Diaspora: The Social Sources of Genetic Genealogy." In *Revisiting Race in a Genomic Age*, edited by Barbara A. Koenig, Sandra Soo-Jin Lee, and Sarah S. Richardson, 253–68. Rutgers Series in Medical Anthropology. New Brunswick, N.J: Rutgers University Press.

———. 2008b. "Bio Science: Genetic Genealogy Testing and the Pursuit of African Ancestry." *Social Studies of Science* 38 (5): 759–83. doi:10.1177/0306312708091929.

———. 2016. *The Social Life of DNA : Race, Reparations, and Reconciliation after the Genome*. Boston, MA: Beacon Press.

Panofsky, Aaron, and Joan Donovan. 2019. "Genetic Ancestry Testing among White Nationalists: From Identity Repair to Citizen Science." *Social Studies of Science* 49 (5): 653–81. doi:10.1177/0306312719861434.

Paschel, Tianna S. 2016. *Becoming Black Political Subjects: Movements and Ethno-Racial Rights in Colombia and Brazil*. Princeton, N.J: Princeton University Press.

Reardon, Jenny. 2011. "Human Population Genomics and the Dilemma of Difference." In *Reframing Rights: Bioconstitutionalism in the Genetic Age*, edited by Sheila Jasanoff, 217–38. Cambridge, Mass.: MIT Press.

Reardon, Jenny, and Kim TallBear. 2012. "'Your DNA Is Our History.'" *Current Anthropology* 53: S233–45. doi:10.1086/662629.

Relethford, John, and Deborah Ann Bolnick. 2018. *Reflections of Our Past: How Human History Is Revealed in Our Genes*. Milton Park, Abingdon, UK; New York, NY: Routledge.

Rosenberg, Noah A., Jonathan K. Pritchard, James L. Weber, Howard M. Cann, Kenneth K. Kidd, Lev A. Zhivotovsky, and Marcus W. Feldman. 2002. "Genetic Structure of Human Populations." *Science* 298 (5602): 2381–385. doi:10.1126/science.1078311.

Royal, Charmaine D., John Novembre, Stephanie M. Fullerton, David B. Goldstein, Jeffrey C. Long, Michael J. Bamshad, and Andrew G. Clark. 2010. "Inferring Genetic Ancestry: Opportunities, Challenges, and Implications." *The American Journal of Human Genetics* 86 (5): 661–673. doi:10.1016/j.ajhg.2010.03.011.

Santos, Ricardo Ventura, Peter H. Fry, Simone Monteiro, Marcos Chor Maio, José Carlos Rodrigues, Luciana Bastos-Rodrigues, and Sérgio D. J. Pena. 2009. "Color, Race, and Genomic Ancestry in Brazil: Dialogues between Anthropology and Genetics." *Current Anthropology* 50 (6): 787–819. doi:10.1086/644532.

Schramm, Katharina, David Skinner, and Richard Rottenburg. 2012. *Identity Politics and the New Genetics: Re/Creating Categories of Difference and Belonging*. New York: Berghahn Books.

Scully, Marc, Steven D. Brown, and Turi King. 2016. "Becoming a Viking: DNA Testing, Genetic Ancestry and Placeholder Identity." *Ethnic and Racial Studies* 39 (2): 162–80. doi:10.1080/01419870.2016.1105991.

Shim, Janet K., Sonia Rab Alam, and Bradley E. Aouizerat. 2018. "Knowing Something versus Feeling Different: The Effects and Non-Effects of Genetic Ancestry on Racial Identity." *New Genetics and Society* 37 (1): 44–66. doi:10.1080/14636778.2018.1430560.

Stoneking, Mark. 2017. *An Introduction to Molecular Anthropology.* Hoboken, New Jersey: Wiley Blackwell.

TallBear, Kim. 2013. *Native American DNA: Tribal Belonging and the False Promise of Genetic Science.* 1st ed. Minneapolis: University of Minnesota Press.

Telles, Edward Eric. 2004. *Race in Another America: The Significance of Skin Color in Brazil.* Princeton, N.J: Princeton University Press.

Wade, Nicholas. 2002. "Gene Study Identifies 5 Main Human Populations, Linking Them to Geography." *New York Times* 20: 36.

———. 2003. "2 Scholarly Articles Diverge On Role of Race in Medicine." *The New York Times*, March 20, 2003, sec. U.S. https://www.nytimes.com/2003/03/20/us/2-scholarly-articles-diverge-on-role-of-race-in-medicine.html.

———. 2014. *A Troublesome Inheritance: Genes, Race, and Human History.* New York: The Penguin Press.

Wagner, J. K. 2010. "Interpreting the Implications of DNA Ancestry Tests." *Perspectives in Biology and Medicine* 53: 231–48. doi:10.1353/pbm.0.0158.

Wagner, Jennifer K., and Kenneth M. Weiss. 2012. "Attitudes on DNA Ancestry Tests." *Human Genetics* 131 (1): 41–56. doi:10.1007/s00439-011-1034-5.

Zhang, Sarah. 2017. "When White Nationalists Get DNA Tests Revealing African Ancestry." *The Atlantic.* August 17, 2017. https://www.theatlantic.com/science/archive/2017/08/white-nationalists-dna-ancestry/537108/.

4 Race, the elephant in the room

Biological anthropological perspectives on race

In previous chapters, we discussed the fundamental concepts of genetic ancestry testing by describing the science and philosophy underlying these tests. Specifically, we introduced what ancestry tests are in Chapter 1. Then, in Chapter 2, we discussed sociocultural approaches to ancestry and explained how embodied difference is one of several ways in which humans think about their communities and the past. In Chapter 3, we examined how genetic ancestry tests are used to reflect upon the past. In those discussions, we alluded to the relationship between race and genetic ancestry but did not extensively discuss what race means in this context. With these fundamental concepts in place, it is now more conducive to think about how genetic ancestry reflects on the present, specifically the sociocultural realities of race and human variation.

As represented by social scientific professional organizations, the general consensus among many social scientists is that human biological variation cannot reasonably be subdivided into biological units otherwise known as race (American Anthropological Association Executive Board 1998; American Sociological Association 2003; AAPA Committee on Diversity 2019). To summarize, these position statements note that compared to other species with similar geographic distributions across the world, humans simply do not exhibit enough difference from each other to be categorized into biological races. Any one or several so-called racial traits, when mapped globally, are non-concordant with what we would expect to find if humans could be divided into discrete racial groups. Furthermore, studies using biological data derived from people across the world have consistently illustrated that there is more variation within local populations or "races" than between populations (Lewontin 1972; Barbujani et al.

1997; Relethford 2002; Mielke, Konigsberg, and Relethford 2010). This means, from a biological standpoint, that there can be more differences between two individuals from the same purported racial group than between two individuals from different purported racial groups. This pattern of within and between group variation undercuts the idea that humans exist in discrete, natural, biologically defined racial groups. In addition, the traits often used to delineate racial groups, for example skin color, hair texture, eye color, or facial features, are non-concordant, meaning that these traits are not uniformly or consistently found in one group to the exclusion of others. Consequently, non-concordance of these types of traits illustrates the arbitrary nature of racial categorization and supports the non-biological nature of race. Additionally, social scientists have noted that racial categories are highly variable across time and geographic space. A now classic study by Marvin Harris (1970) illustrates just how variable racial categorizations can be. Harris examined racial categorizations in Brazil by showing 72 portraits featuring differing combinations of one of three skin colors, one of three hair textures, one of two lip shapes, and one of two types of noses. Based on the responses of 100 study participants, Harris found that racial categorizations in Brazil were very different than the categorizations found in the United States. The respondents provided 492 different categorizations of the faces they were shown. Harris's cross-cultural study of racial categorizations support the idea that race is not biological in nature, but rather is more instructive about the variety of ways that humans organize and give value to human difference.

Similar conclusions may be drawn by looking at how race categories have disappeared, appeared, and been renamed in the US census in response to social, economic, and political circumstances (Lee 1993). For example, in the 1960 United States census, Hawaiian and Aleut are listed for the first time in the census. This change in the census does not represent the emergence of new racial groups or some sort of change in human biology, but rather marks a political change in the United States, as Hawaii and Alaska became states in 1959. All of these examples indicate that race itself is not simply biology but rather something that people construct as they make sense of the world around them. Race is a product of culture, and though it may be intimately tied to biology, race is not biology.

Biocultural approaches to racial experience

Despite the recognition that there is no biological basis to race, social scientists are also cognizant that race can affect biology.

Within anthropology and related fields, the concept of how cultural phenomenon is manifested in the body is referred to as embodiment (Csordas 1990; Mascia-Lees 2011; Meier et al. 2012). Numerous researchers have relied on the concept of embodiment to explore how social constructs such as race and gender can influence biology (Krieger et al. 2005; Fausto-Sterling 2008; Gravlee 2009; Kuzawa and Sweet 2009; Lee 2014). These approaches are similar to how we explained race as a form of embodied difference in the previous chapter. These studies have proven insightful in illustrating how people's ideas about race shape how individuals interact with one another and how these ideas can influence the types of resources that are, or are not, accessible. Inequitable access to education, healthcare, quality food, economic opportunity, and other resources are influential on an individual's overall health and well-being. Investigations into the biological mechanisms of how race actually gets "under the skin" are ongoing. Many of these types of studies have specifically focused on stress in relation to discrimination as the leading mediating element of how race becomes embodied and affects health outcomes (Davis 2005; Djuric et al. 2008; Surtees et al. 2011; Rodney and Mulligan 2014; Berger and Sarnyai 2015; Ridout, Khan, and Ridout 2018). The findings of such studies have not yet been entirely conclusive, but they do present some promising ideas about how biological functions related to cellular aging and behavior are affected by adverse events including stress and racism. When we think about the embodiment of difference, we are committed to exploring both the biological and cultural dimensions of such embodiment; in fact, this is how we frame our biocultural approach.

Biocultural refers to an approach to understanding human experience that is sensitive to the dynamic interactions of biological and cultural factors that shape human experience (Goodman and Leatherman 1998; Dufour 2006; Thomas 2016; Zuckerman and Martin 2016). As we apply this same type of approach to thinking about the relationship between race and genetic ancestry, we advocate an understanding of race that is consistent with, yet goes beyond, the "race is a social construct" paradigm. Rather, we operationalize race as experience where there is an embodied, sensed, collective, bodily distinction that is inclusive of both positive and negative experiences (Benn Torres and Torres Colón 2015). Here, "racial experience" attends to the reality of biological diversity as well as the experiences that emanate from that biological diversity. Racial experience shapes how people view themselves and other people and this understanding of self and others extends to senses of community. It is also important to note that in our formulation, we go further than most anthropologists who study racial embodiment when we account for the "good and bad" experiences. Racial experience is inclusive of racism, but it is also much more than racism. For example, there are many aesthetic social behaviors that are racially experienced without being experienced as racist (Torres Colón 2018). Other racial experiences, such as political activism, respond to racism in the world while fomenting positive ideas about racial groups (e.g., "Brown power" or "Black is beautiful"). Therefore, while racism can get into the body in the form of stress that then defines a racial group, positive racial experiences can also serve as a buffer to racism and lead to resilience in the survival of racialized communities (Mullings and Wali 2001; Benn Torres and Torres Colón 2015; Hunter et al. 2016).

Operationalizing race as the full range of racial experience compels us to recognize social and biological realities as part of the same ontological continuum. In doing this, we assert the importance of both the social realities of race as well as the reality of human biological diversity. Utilizing racial experience as the fundamental approach to race, we intend to uphold the anti-racist traditions that mark contemporary anthropology and be attentive to human experience and variation. That is to say, as we advocate this framework for thinking about race, it is important to note that recognizing human variation is not in opposition to anti-racist stances. Rather, we remain decisive in the non-biological basis to race yet are sensitive to the social dimensions of race and racism that shape the everyday lives of people.

Genetic ancestry and race

With regards to the relationship between genetic ancestry and race, operationalizing race as racial experience is instructive for understanding how genetic data may or may not play some role in shaping a test-taker's ideas about race. As discussed in Chapter 1, genetic ancestry tests query the past using biological data and can be informative about the geographic origins of a test-taker's ancestors. Furthermore, as discussed in Chapters 2 and 3, we covered how there are many ways in which genetic ancestry test results can be interpreted and that the motivations behind how test-takers approach genetic ancestry and the ensuing results are subject to historic, economic, and political factors. The same complexities are applicable to thinking about how genetic ancestry relates to race. From a practical standpoint, genetic ancestry cannot define a racial group for all of the reasons why race is not a biologically valid category. Despite the reality of being able to cluster groups of people due to global patterns of population substructure, the ways that people self-identify varies within these clusters as a result of social convention, historical circumstances, and other relevant factors such as economics or politics. This disconnect between ancestry results and how people self-identify in general makes equating genetic ancestry with specific self-identified racial groups extremely problematic.

This point is nicely illustrated in the results of a large-scale study conducted by researchers working with genetic ancestry results from about 160,000 23&Me customers (Bryc et al. 2015). As part of this study, test results were considered in relation to customers ethnic and racial self-identification. Bryc and colleagues report that in general, participants with less than 28% African ancestry tended to self-identify as European American, however, some study participants with as little as 10% African ancestry self-identified as African American. Furthermore, Bryc and colleagues found that only 3.5% self-identified European Americans have at least 1% African ancestry, meaning the vast majority, nearly 97%, of European Americans do not have African ancestry. Considered in its entirety, this means, that within clusters associated with self-identified European Americans, there is primarily European ancestry; however, within clusters associated with self-identified African Americans, there are individuals who have substantial amounts of European ancestry yet still self-identify as African American. Stated differently, despite having substantial European genetic ancestry, there are people that still self-identify as African American, yet individuals that have substantial African ancestry virtually never self-identify as European American. This extensive variability of African ancestry among self-identified African Americans may be attributed to social convention and reflects the constructs created and maintained by contemporary people around the idea of who is and is not considered white (Harris 1964; Hickman 1996). Nevertheless, Bryc and colleagues conclude that,

> Contrary to expectations under a social one-drop rule, or "Rule of Hypodescent," which would mandate that individuals who knowingly carry African ancestry identify as African American, the probability of self-reporting as African American given a proportion of African ancestry follows a logistic probability curve, suggesting that individuals identify roughly with the majority of their genetic ancestry.
>
> (Bryc et al. 2015, 50)

Though Bryc and colleagues entertain the possibility of sampling error, their interpretations are certainly supported by their data.

However, one must also recognize the implications of the fact that the vast majority of European Americans only have European ancestry while the same is not true for African Americans. This ancestry pattern shows a remarkable commitment to the status quo of how people self-identify in relation to the Rule of Hypodescent. In this example, we only referenced African and European Americans, when Latinos, also part of Bryc et al.'s study, are added to this conversation, the relationship between genetic ancestry and how people self-identify becomes more complex. In the case of Latinos, the relationship between self-identified race or ethnicity and genetic ancestry is complicated due to a number of factors. These factors include, but are not limited to, very different histories and different social conventions surrounding how people identify which ultimately can affect mate choice. As a brief example, Latinos in the study included people that self-identified as coming from the Hispanic Caribbean as well as from Mexico, and Central and South America. Interestingly, the ancestral origins varied substantially within and between these self-identified populations, yet they were still broadly categorized as Latino. What the Bryc et al. study illustrates is that for many people within the United States, there are no absolutes between self-identified race and genetic ancestry.

Another factor to consider when thinking about race and genetic ancestry is the dynamic nature of race. As mentioned earlier in this chapter, within the United States census, racial categories appear, are rebranded, or even disappear and this is reflective of changing social, cultural, and political contexts. In addition, despite common conceptions that race is an unchanging characteristic, social scientists have shown that race is not actually as immutable as one might expect (Brodkin 1998; Goodman 2000, 2017; Ajrouch and Jamal 2007; Loveman and Muniz 2007; Ignatiev 2012). These studies highlight how sometimes due to bureaucratic factors, fluctuating public sentiment related to political and economic changes, among other things, that people can be born into one race and later die as a member of a different race. In addition, the articulations of race vary substantially across geographies. Harris's 1970 study, discussed earlier in this chapter, neatly illustrates how ideas about which factors are important in constructing racial groups do not translate or are not applicable across different communities. These examples highlight the social nature of race and how experience shapes how individuals self-identify and are identified. Furthermore, it is possible that a given test-taker has individuals in their genetic genealogy that, if living today, would self-identify in a different manner than the test-taker. Our own ancestry results illustrate this point nicely. Both of us have some degree of

European ancestry. However, neither of us self-identify as European American. Jada identifies as Afro-Caribbean/African American and Gabriel as Puerto Rican/Latino. From the standpoint of genetics, we both recognize that what we have inherited from all of our ancestors is important in making us who we are. However, we also recognize that the ways in which we self-identify goes well beyond the results of an ancestry test. Rather, the ways in which we self-identify emanates from our socialization and experiences within the Caribbean and the United States. Our example is hardly unique, as evidenced by other researchers including the Bryc's et al.'s (2015) study mentioned earlier as well as the experiences of the lay public like those shown in the momondo's *The DNA Journey* discussed in Chapter 3. It is because of the dynamic, social nature of racial experience, that genetic ancestry cannot be equated to race. Simply put, genetic ancestry tests cannot tell you your race or ethnicity, but it can tell you something about the geographic origins of many of your ancestors. For historical and cultural reasons, geographic origins are often thought about as associated with specific racial groups, which can lead people to falsely equate genetic ancestry with race. Ultimately, how people self-identify is a dynamic and complex social process that changes in response to time and space. Accordingly, no genetic ancestry test can capture the nuances involved in identity construction.

Genetic ancestry as racial experience

Anthropologists carry out their work in social contexts, and they often position the relevance of their research in response to social problems. Historically, anthropologists have developed the three-part response to the social problem of race. First, race is not biology. Second, race is cultural in nature. Third, there is a biocultural insight, which tells us that people experience race in cultural ways that have biological consequences. However, we also suggest that the "cultural ways" in question can be positive, not just racism. While carrying out genetic ancestry and ethnographic fieldwork, we have encountered a conundrum that, at first, seems to escape the explanatory power of anthropological approaches to race.

The conundrum is as follows. People encounter anthropological work on their history, culture, and ancestry (carried out by us)—not as anthropological insights—but as racial experience. Specifically, people in different societies experience knowledge about their genetic ancestry in ways that are meaningful in relation to their racialized societies and cultures. Rather than simply instill or fortify a biological notion of race, genetic ancestry resonates with multiple ways in which people experience race. Now, occasionally we have encountered people who subscribe to scientific ideas about race, and they interpret genetic ancestry as an extension of those (pseudo)scientific ideas. However, the vast majority of people who we have worked with throughout the Caribbean and the United States experience genetic ancestry information in ways that are most consistent with political, aesthetic, and historical ideas that emerge from their collective experiences of race. This presents a conundrum to social scientists who want to declare that there is no relationship between biology and race, for if we take biocultural insights about race seriously, a biological fact—such as genetic ancestry—can indeed be bioculturally experienced as racial experience. Of course, the conclusion is not that race is biologically determined, but that the science about human diversity can be racially experienced. Furthermore, telling people that there is no relationship between biology and race might work well enough in university class-rooms, but such lessons may be pretty irrelevant in a world where racial experience shapes most understanding of human biological and cultural diversity.

During the course of our work in another ancestry project, this time with the Santa Rosa First Peoples Community who reside in northern Trinidad, we had many conversations with research participants about their "mixed" background. They had full knowledge that they had African ancestors along with First Peoples ancestors. The question of hybridity did not lead to doubts about their authenticity as descendants from indigenous ancestors. Instead, they group their African and indigenous ancestors into their collective identity of First Peoples. Several research participants were very specific about knowing of their "hybridity," and how this "hybridity" in no way negated their sense of pride in being "Carib." Part of the reason that many members from the community chose to participate in the genetic ancestry project was because they wanted to learn more about their ancestors, and specifically their indigenous ancestors. They demonstrated no fear in what the genetic ancestry might reveal, and we have found much enthusiasm through the duration of the project and leading up to Jada's presentation during the 2017 First People's Day in Trinidad, where much of the research was discussed before the community and a large audience of academics and the general populace.

Contrary to what at least one uninformed and ill-natured commentator has said about the intent of our genetic ancestry work in Santa Rosa (Forte 2013), our research question concerned the community's population history, and "proving" indigenous identity through genetic testing was never considered. The people in the community who participated in the research brought a diverse set of perspectives regarding genetic technology to the project. Most did not really have a good sense of the processes through which ancestry data is ascertained from cheek swabs. They did, however, have a clear sense that the research results would comment on a group of indigenous ancestors that lived in Trinidad and northern South America at the time of European colonization. Their sense of these ancestors was specific and differentiated from other ancestors, mainly the African and South Asian ancestors that compose the more recent populations of Trinidad. In community meetings before and after returning results, it became patently clear to us that many people in Santa Rosa felt collective unity due to having a group of indigenous ancestors that survived through colonialism and flourished through the hybridity that resulted from that colonial process. Genetic ancestry was a vehicle through which they could experience what they had been experiencing already: being indigenous. Genetic ancestry was a new way for us, as anthropologists, to see the population history of the Caribbean. For participants, it was an innovative way for them to reassert their identity.

The use of genetic ancestry as a way. to reassert a collective identity has resonated in our work throughout the Caribbean. Garifuna in St. Vincent have similarly embraced collaborative genetic ancestry research projects. There, notions of hybridity are fully intelligible with racial groups. Similarly in Dominica, where the Kalinago community chose not to participate in genetic ancestry research projects, we talked to quite a few people who joked about how obvious it was that there was a lot of mixing in their families, yet their indigeneity survived. People's collective sensibilities about hybridity are a form of collective bodily distinction. This collective sense of distinction is particularly resistant to the Black-Indigenous dichotomy that is so firmly held throughout the Americas, but it nonetheless constitutes a form of racial experience. Genetic ancestry, whether embraced or not by these communities, has been a way for people to experience their sense of the past and the collective bodies that make up the ancestors in the past. How much value these communities chose to make out of their ancestry results has varied. In Trinidad, for example, despite Jada's repeated statements about how genetic ancestry cannot tell anyone who is or is not indigenous, we often heard leaders declare that her study demonstrated that they were indigenous.

These statements about how genetic ancestry "proves identity" are often out of the control of researchers. However, rather than seeing their misuse as inherent in genetic technology, the so-called misuse is an extension of how people think about and manage their racial identities in the real world. The Santa Rosa First Peoples Community had been engaged in a long struggle to gain recognition from a Trinidadian government that doubted their indigeneity. Part of the reason that government officials doubted their indigeneity is because they perceived the community to be too mixed and lacking the absence of authentic "Indians" like one would see on a TV documentary about the Amazons. In other words, the government doubted the racial authenticity of the community. Members of the Santa Rosa community and those outside of it had developed two different cultural notions of what hybridity meant for racial authenticity. However, no one had stopped to question the validity of race as a biological category—it simply was not an issue. So when the genetic ancestry projects arrive, there is nothing to reify that is not already there. Moreover, in the face of ancestry results that show mixed biogeographic ancestry, the people of the Santa Rosa First People's Community did not reconsider their stance of hybridity. Why would they? The genetic results validated and were a tangible experience of how they understood race in Trinidad.

Having considered the biocultural nature of race and racial experiences of genetic ancestry, we now consider the intellectual history of race within anthropology. This discussion will help to situate genetic ancestry within a broader framework of anthropological inquiry in which genetic ancestry can be utilized in ways that align with the anti-racist traditions that characterize contemporary biological anthropology.

References

AAPA Committee on Diversity. 2019. "AAPA Statement on Race & Racism." http://physanth.org/about/position-statements/aapa-statement-race-and-racism-2019/.

Ajrouch, Kristine J., and Amaney Jamal. 2007. "Assimilating to a White Identity: The Case of Arab Americans1." *International Migration Review* 41 (4): 860–79. doi:10.1111/j.1747-7379.2007.00103.x.

American Anthropological Association Executive Board. 1998. "American Anthropological Association Statement on Race." *American Anthropologist* 100 (3): 712–13.

American Sociological Association. 2003. *Statement of the American Sociological Association on the Importance of Collecting Data and Doing Social Scientific Research on Race.* Washington, DC: American Sociological Association.

Barbujani, Guido, Arianna Magagni, Eric Minch, and L. Luca Cavalli-Sforza. 1997. "An Apportionment of Human DNA Diversity." *Proceedings of the National Academy of Sciences* 94 (9): 4516–519. doi:10.1073/pnas.94.9.4516.

Benn Torres, Jada, and Gabriel A. Torres Colón. 2015. "Racial Experience as an Alternative Operationalization of Race." *Human Biology* 87 (October): 306–12.

Berger, Maximus, and Zoltán Sarnyai. 2015. "'More than Skin Deep': Stress Neurobiology and Mental Health Consequences of Racial Discrimination." *Stress* 18: 1–10.

Brodkin, Karen. 1998. *How Jews Became White Folks and What That Says about Race in America.* New Brunswick, NJ: Rutgers University Press.

Bryc, K., E. Y. Durand, J. M. Macpherson, D. Reich, and J. L. Mountain. 2015. "The Genetic Ancestry of African Americans, Latinos, and European Americans across the United States." *American Journal of Human Genetics* 96: 37–53. doi:10.1016/j.ajhg.2014.11.010.

Csordas, Thomas J. 1990. "Embodiment as a Paradigm for Anthropology." *Ethos* 18 (1): 5–47. https://www.jstor.org/stable/640395.

Davis, S. K. 2005. "Stress-Related Racial Discrimination and Hypertension Likelihood in a Population-Based Sample of African Americans in the Metro Atlanta Heart Disease Study." *Ethnicity and Disease* 15: 585–93.

Djuric, Z., C. E. Bird, A. Furumoto-Dawson, G. H. Rauscher, M. T. Ruffin, R. P. Stowe, K. L. Tucker, et al. 2008. "Biomarkers of Psychological Stress in Health Disparities Research." *The Open Biomarkers Journal* 1: 7–19. doi: 10.2174/1875318300801010007.

Dufour, Darna L. 2006. "Biocultural Approaches in Human Biology." *American Journal of Human Biology* 18 (1): 1–9. doi:10.1002/ajhb.20463.

Fausto-Sterling, A. 2008. "The Bare Bones of Race." *Social Studies of Science* 38: 657–94. doi:10.1177/0306312708091925.

Forte, Maximilian C. 2013. "Who Is an Indian? Race, Place, and the Politics of Indigeneity in the Americas." *Zero Anthropology* (blog). August 23, 2013. https://zeroanthropology.net/2013/08/23/who-is-an-indian-race-place-and-the-politics-of-indigeneity-in-the-americas/.

Goodman, Alan H. 2000. "Why Genes Don't Count (for Racial Differences in Health)." *American Journal of Public Health* 90 (11): 1699–702. https://www.ncbi.nlm.nih.gov/pmc/articles/PMC1446406/.

———. 2017. "Reflections on 'Race' in Science and Society in the United States." *Journal of Anthropological Sciences* 95: 283–90. doi:10.4436/JASS.95008.

Goodman, Alan H., and Thomas L. Leatherman. 1998. *Building a New Biocultural Synthesis Political-Economic Perspectives on Human Biology.* Linking Levels of Analysis. Ann Arbor: University of Michigan Press.

Gravlee, C. C. 2009. "How Race Becomes Biology: Embodiment of Social Inequality." *American Journal of Physical Anthropology* 139: 47–57. doi:10.1002/ajpa.20983.

Harris, Marvin. 1964. *Patterns of Race in the Americas.* Vol. 1. Book, Whole. New York: Walker.

———. 1970. "Referential Ambiguity in the Calculus of Brazilian Racial Identity." *Southwestern Journal of Anthropology* 26 (1): 1–14. http://www.jstor.org/stable/3629265.

Hickman, Christine B. 1996. "The Devil and the One Drop Rule: Racial Categories, African Americans, and the U.S. Census." *Michigan Law Review* 95: 1161–265. https://heinonline.org/HOL/P?h=hein.journals/mlr95&i=1182.

Hunter, Marcus Anthony, Mary Pattillo, Zandria F. Robinson, and Keeanga-Yamahtta Taylor. 2016. "Black Placemaking: Celebration, Play, and Poetry." *Theory, Culture & Society* 33 (7–8): 31–56. doi:10.1177/0263276416635259.

Ignatiev, Noel. 2012. *How the Irish Became White*. New York: Routledge. doi:10.4324/9780203473009.

Krieger, Nancy, Kevin Smith, Deepa Naishadham, Cathy Hartman, and Elizabeth M. Barbeau. 2005. "Experiences of Discrimination: Validity and Reliability of a Self-Report Measure for Population Health Research on Racism and Health." *Social Science & Medicine* 61 (7): 1576–596. doi:10.1016/j.socscimed.2005.03.006.

Kuzawa, Christopher W., and Elizabeth Sweet. 2009. "Epigenetics and the Embodiment of Race: Developmental Origins of US Racial Disparities in Cardiovascular Health." *American Journal of Human Biology* 21: 2–15. doi:10.1002/ajhb.20822.

Lee, Emily S. 2014. *Living Alterities: Phenomenology, Embodiment, and Race*. Albany, New York: SUNY Press.

Lee, Sharon M. 1993. "Racial Classifications in the US Census: 1890–1990." *Ethnic and Racial Studies* 16: 75–94.

Lewontin, Richard C. 1972. "The Apportionment of Human Diversity." *Evolutionary Biology* 6: 381–98.

Loveman, Mara, and Jeronimo O. Muniz. 2007. "How Puerto Rico Became White: Boundary Dynamics and Intercensus Racial Reclassification." *American Sociological Review* 72 (6): 915–39. doi:10.1177/000312240707200604.

Mascia-Lees, Frances E., ed. 2011. *A Companion to the Anthropology of the Body and Embodiment*. Vol. 22. Blackwell Companions to Anthropology 13. Chichester; Malden, MA: Wiley-Blackwell.

Meier, Brian P., Simone Schnall, Norbert Schwarz, and John A. Bargh. 2012. "Embodiment in Social Psychology." *Topics in Cognitive Science* 4 (4): 705–16. doi:10.1111/j.1756-8765.2012.01212.x.

Mielke, James H., Lyle W. Konigsberg, and John H. Relethford. 2010. *Human Biological Variation*. Vol. 2. New York: Oxford University Press.

Mullings, Leith, and Alaka Wali. 2001. *Stress and Resilience: The Social Context of Reproduction in Central Harlem*. New York: Kluwer Academic/Plenum Publishers.

Relethford, J. H. 2002. "Apportionment of Global Human Genetic Diversity Based on Craniometrics and Skin Color." *American Journal of Physical Anthropology* 118: 393–98.

Ridout, Kathryn K., Mariam Khan, and Samuel J. Ridout. 2018. "Adverse Childhood Experiences Run Deep: Toxic Early Life Stress, Telomeres, and

Mitochondrial DNA Copy Number, the Biological Markers of Cumulative Stress." *BioEssays* 40 (9): 1800077. doi:10.1002/bies.201800077.

Rodney, Nicole C., and Connie J. Mulligan. 2014. "A Biocultural Study of the Effects of Maternal Stress on Mother and Newborn Health in the Democratic Republic of Congo." *American Journal of Physical Anthropology* 155: 200–09.

Surtees, P. G., N. W. Wainwright, K. A. Pooley, R. N. Luben, K. T. Khaw, D. F. Easton, and A. M. Dunning. 2011. "Life Stress, Emotional Health, and Mean Telomere Length in the European Prospective Investigation into Cancer (EPIC)-Norfolk Population Study." *The Journals of Gerontology. Series A, Biological Sciences and Medical Sciences* 66: 1152–162. doi:10.1093/gerona/glr112.

Thomas, R. Brooke. 2016. "Exploring Biocultural Concepts: Anthropology for the Next Generation." In *New Directions in Biocultural Anthropology*, edited by Molly K. Zuckerman, and Debra L. Martin, 29–47. Hoboken, New Jersey: Wiley-Blackwell. doi:10.1002/9781118962954.ch2.

Torres Colón, Gabriel A. 2018. "Racial Experience as Bioculturally Embodied Difference and Political Possibilities for Resisting Racism." *The Pluralist* 13 (1): 131–42. https://www.jstor.org/stable/10.5406/pluralist.13.1.0131.

Zuckerman, Molly K., and Debra L. Martin. 2016. *New Directions in Biocultural Anthropology*. Hoboken, New Jersey: Wiley-Blackwell.

5 Anténor Firmin and biological anthropology as anti-racism

If how a genetic ancestry test shapes people's understanding of themselves and others is relative to cultural notions of the past and relatedness, then the study of human similarities and differences—i.e., anthropology—can similarly be understood in the cultural context of its practitioners. In Chapter 3, we challenged the views of scholars whose critique presumes that genetic ancestry has a negative impact on society by reinforcing racial categories. However, our challenge of these perspectives extends beyond different interpretations of how humans interact with genetic ancestry. Instead, we presented a broader view on how to think about race—scientifically, as a social problem, and a political question. At the heart of this perspective is the role of anthropology as a social and biological science in shaping anti-racist thought, and vice versa. Genetic ancestry technology, as a scientific tool and a cultural form, is increasingly at the center of conversations about race and racism, yet modern science, race, and racism have been deeply intertwined since the emergence of anthropology (Marks 2009). Consequently, genetic ancestry is just one of the latest scientific technologies to be entangled with race. In this chapter, a brief theoretical and historical examination of the intersections between race, science, and anthropological thought will be useful to more fully understand how genetic technology can serve in an anti-racist anthropology. Taking seriously the lessons about racial experience discussed in the previous chapter, we build on those lessons to illustrate what science looks like when articulated as an extension of racial experience.

In this chapter, we specifically focus on the intellectual heritage of Haitian anthropologist Anténor Firmin and other anti-racist intellectuals that challenged biological notions of race when anthropology as a professional field was still young. In doing this we seek to link early anthropological perspectives on human variation to contemporary ideas about the experiential nature of race, including genetic ancestry. In the previous chapters, we explored how sociocultural context matters when interpreting genetic ancestry—essentially, one's sociocultural background can affect how one engages science as a tool for learning about humans. In this chapter, we are applying that analysis of sociocultural context to anthropologists. That is, we examine how sociocultural context affects how scientists think about relatedness and ancestry. Because early anthropology emerged and was immersed in colonialism and racism, the greater purpose of this chapter is to examine the role of biological anthropology, and biocultural theory more specifically, in the study of race. After a careful reading of Anténor Firmin, we examine how he can help us think about race and biological anthropology today—and specifically in relation to two other important anti-racist intellectuals for anthropology, W. E. B. Du Bois and Franz Boas. We then argue for scientific practices that fully incorporate anti-racism as a best scientific practice. Interviews with biological anthropologists from racialized communities in the United States, which highlight how their racialized experiences sharpen their scientific focus, help substantiate our overall argument for this chapter. We believe there are valuable lessons in intellectual history because we often think that contemporary critics of scientific racism (i.e., anthropologists) confuse their anti-racist scholarship with a holistic understanding of how race operates in the world. Moreover, intellectual historical accounts can go beyond a genealogical exercise. The lessons learned from studying how theory has been developed in sociocultural context can be used to inform contemporary theory. With a solid understanding of anthropological anti-racist intellectual histories, as anthropologists we can better understand how our theorizing is shaped by our own sociocultural contexts.

Anténor Firmin and the equality of human races

In 1885, Anténor Firmin published *The Equality of the Human Races* (2000), which is a monumental critique of racist anthropology. Franz Boas's 1911 *Mind of Primitive Man*, which is one of the most important works of early American anthropology, was of great importance to anti-racist intellectual thought in the United States. *Equality* foreshadows the *Mind of Primitive Man*'s method of inquiry—dismantling faulty evolutionary biology that justified racial inequality, exposing the history of great civilizations around the world to argue for the equality of races, and aspiring for the proper cultural and political conditions where all races can flourish. Although Firmin had less biological anthropological data, he managed to produce an elegant scientific argument inspired by an anti-colonial spirit, 26 years earlier than Boas's now foundational work. This seminal achievement was not made possible by the anti-racist application of anthropological ideas, but by a black Haitian intellectual who seamlessly aligned biological anthropology and anti-racism. Anthropology was anti-racist before it was anthropology; and if anthropologists claim Boas as the father of American anthropology for the four-field approach that is exemplary exhibited in *Mind of Primitive Man*, then anthropologists might have a long-lost grandfather in Haiti.

Firmin was born in Cap-Haitian, located on the northern coast of Haiti, on November 27, 1850. Cut from the same mold as other Caribbean intellectuals of his time, Firmin was broadly educated in the humanities, sciences, and law. He held political appointments in Haiti and was sent as a diplomat to Paris in 1883. He was admitted to the Paris Anthropological Society in 1884. Once in the Society, Firmin opted to write *Equality* rather than initiate an open debate about the equality of races. His motives were clear from the beginning of the book:

> I do not have to hide it. I am always shocked whenever I come across dogmatic assertions of the inequality of the races and the inferiority of Blacks in various books. Now that I have become a member of the *Société d' anthropologie de Paris*, such assertions seem to me even more incomprehensible and illogical.
>
> (2000, liv)

Not having much time in Paris before having to return to Haiti, Firmin finds himself in a hurry to write a book that he might not be able to finish in the future,

> The haste with which I undertook the project undoubtedly has affected its execution … Time was of the essence, and I was not sure that any of my Black colleagues had both the good will and the patience on needs in order to construct, combine, and present the arguments and the research materials in the way I strove to do.
>
> (2000, liv)

Although Firmin does not consider himself a scientist or a hero, and he makes his intellectual perspective crystal clear before embarking on his anthropological argument for the equality of all races, "I am Black. Moreover, I have always considered the religion of science as the only true one, the only one worthy of the attention and infinite devotion of any man who is guided by reason" (Firmin 2000, lv). His blackness is intellectual grounds from where he finds utility for science as a religion of reason. Furthermore, Firmin dedicates the book to the "Black race": "This book is a humble and respectful offering I make to the race in a religious spirit. Others will do better than I someday, but no one will ever hope more for its regeneration and wish for its glory than I do" (2000, lix).

After making the case for a positivist/scientific anthropology (*Equality*'s subtitle is, in fact, "Anthropologie Positive") and reviewing basic classificatory tenets of evolutionary biology, Firmin aims to quickly dismantle polygenism as an empirical domain from which to rank order human races. In the absence of scientific evidence demonstrating the common origin of humans, Firmin takes two general argumentative strategies. First, he establishes, without equivocation, the equal worth of the "Black race" with that of the "White race"; and he follows that argument with one that resonates with the philosophical origins of modern anthropology. Firmin quotes M. Flourens: "The unity of intelligence is the ultimate and definitive proof of the unity of humanity" (1861). Therefore, if all the human races share the same mental and spiritual capacity, polygenism seems unlikely and—even if true—irrelevant for rank ordering races. A polygenist argument that places Europeans at the top of the hierarchy is not possible if it is established that all human races are equal.

Second, in several chapters, Firmin argues for what today we call the non-concordance of racial traits (see the Chapter 4 section titled "Biological anthropological perspectives on race"), with a lengthy initial emphasis on craniology. Firmin argues that when anthropometric evidence is closely scrutinized, there is too much variation or overlapping variation to logically create a human racial taxonomy. Firmin is attentive to the available scientific evidence from several anthropological schools of thought. Like Franz Boas would later argue, Firmin also commits a large portion of his book to a comparative ethnological exercise in order to demonstrate the psychological equality of races. With regard to language, for example, he concludes, "... languages, in their very essence, have more to do with the nature of a society, of civilization, than with race" (Firmin 2000, 133). In measuring "aptitude," Firmin offers historical evidence from Egyptian mathematics and contemporary abilities of Black Haitians that the Black race is equally capable.

But perhaps the most interesting argument about the non-concordance of racial traits is about "beauty in the different human races" (Firmin 2000, 181–200). Firmin attempts to provide a logical criteria for evaluating beauty. In Chapter 4, we discussed the many sociocultural variables that differently shape racial experience. Firmin's scientific engagement with beauty is an interesting example of how scientific practice is always entangled in sociocultural context. He provides examples of beautiful and ugly people for different races in order to dismiss the notion that Europeans are always the most beautiful and Black people are always the ugliest. He then presents a curious argument about environmental plasticity and historical circumstance that can be summarized as follows: many Black peoples in Africa live in "backward" conditions so they appear "ugly," but given the proper conditions of progress and achievement of civilization, then "beautiful" people will too emerge in Africa. He accentuates these arguments with a discussion of beautiful people in Haiti. Putting aside the contemporary implausibility of Firmin's empirical approach to beauty, it is important to understand that although he accepts the ugliness of some Black people, he is rejecting the heritability of ugliness vis-a-vis the Black race. His argument is not against race, it is against the inequality of races and racist science.

In addition to anti-racist scientific arguments, Firmin also argues against the scientific bias of many anthropologists (Beckett 2017, 170–72). Against the leading anthropological proponents of racial ranking, Firmin is always careful to recognize the brilliance of "the scientist," their methods, and their scholarship; but he is critical of their arguments. Their science is not at fault, the problem is that scientists of his time, and specifically the anthropologists of the Paris Anthropological Society, had already made up their minds against the equality of the races. Firmin sees nothing unique about the historical degradation of the Black race. Instead, he traces the ranking of races to economic interest, particularly of slavers; and he uses the Roman treatment of slaves as an example (Firmin 2000, 139–44). It is crucial to recognize how Firmin chooses to engage the problem of racism in this regard.

> Certainly, if anthropologists did grasp all the data needed to do good anthropology, no one would be better qualified nor more authorized than they to handle questions in this field. Unfortunately, despite the relative independence science has acquired in this our century of freedom, it is still influenced by ambient ideas. It suffices that a very talented scientist, capable of taking the leadership of a scientific current, adopts one of these attractive but ephemeral ideas and clothes it in the respectable garb of specific formulae and methods, and a school emerges, blocking progress in this particular branch of science. And so it goes until

it becomes obvious that the great man of science had erred. At this point, investigations are conducted, discussions are held, and the truth is vaguely intuited. Sometimes involved scientists reason so rigorously that they come on the verge of proclaiming the truth. But there's the rub! If this truth contradicts the school's official position, if it is contrary to the master's word, then these scientists will suddenly become inarticulate rather than stand against the official theory.

(Firmin 2000, 144–45)

This position on science leads Firmin to careful and balanced dismissals of racial differences with regard to intelligence. He cites the lack of evidence supporting racial disparities in intelligence and details the lack of scientific knowledge about the brain that one could use to test racial differences (Firmin 2000, 162–63). One can sense Firmin's deep admiration for science, and how he ties scientific progress to the Enlightenment and the French Revolution.

It is so beautiful to see human intelligence soar to the most abstract conceptions and project brilliant flashes on the dark history of past eras as man tries to lift the veil that covers the secrets of nature as it did the ancient Isis. On the other hand, it is no less reasonable to acknowledge the frequent failure of man's noble efforts. As soon as we reach the greatest heights of science, we are surrounded by an atmosphere of doubt and discouragement that would be sufficient to destroy the human spirit.

(Firmin 2000, 82)

We can see how the polluting of science and anthropology with racist ideas represents a greater assault on the "human spirit." His corrective and anti-racist science emerges from an intellectual and spiritual space of Black freedom, Black liberation, and the improvement of the Black race (see Charles 2014 for a similar geopolitical interpretation).

This intellectual space of freedom, liberation, and improvement is rebellious and anti-colonial. One of the most gratifying aspects of reading *Equality* is the many instances when Firmin poetically affirms freedom, equality, and self-worth. For example, his discussion on monogenism and polygenism begins with the following assertion:

Whether the human species is one or many, my thesis will be no more or less difficult. It does not matter to me at all that some people consider the Black race, my race, a distinct species from the White, Yellow, and Red races, or from the sixteen different categories based on the specific colors and shapes identified by the whimsical imagination of the polygenists. Whatever taxonomic distance may separate my race from all others, I know that it holds an unchallengeable place in the world and has nothing to envy in any other race in terms of intelligence, virtue, and will power.

(Firmin 2000, 35–6)

Later in the book and while dismantling craniological justifications for racial inequality, he takes time to justify his scientific effort by articulating his greater purpose,

Motivated by an insatiable thirst for the truth and the obligation to contribute, no matter how modestly, to the scientific rehabilitation of the Black race whose pure and invigorating blood flows in my veins, I take immense pleasure in navigating through these columns of figures arranged with such neatness for the edification of the intellect.

(Firmin 2000, 102)

Indeed, by the end of *Equality*, Firmin makes the case for Black liberation through the example of Haitian liberation. In short, he argues that given the right conditions of freedom and modernity, Black people would one day be equal to Europeans and, even more, "play a dominant role in world history" (Firmin 2000, 445).

Claiming Firmin as an intellectual ancestor

Anténor Firmin was not trained as an anthropologist, but few intellectuals who are today considered foundational to the discipline (e.g., Lewis H. Morgan or Edward B. Tylor) received formal anthropological training because anthropology was not yet institutionalized in academia. *Equality* engages anthropological research by analyzing and reanalyzing research from biological and historical perspectives. In addition, Firmin provides his own empirical examples as part of his commentary on humanity. By these criteria, *Equality* is similar in form to other anthropological texts of its era, and anthropologists today could easily reclaim Firmin as a lost intellectual ancestor. Yet Firmin lacks presence and influence in the anthropological canon. His influence, then, must be projected into the present, and we have to make conscious decisions as to why he is important as an intellectual ancestor and what, if any, is the theoretical significance of his work and sociocultural context in which he practiced anthropology. As applied to our examination of the broader cultural meanings and social contexts of genetic ancestry, we can reclaim Firmin through engaging in a cross-cultural comparison of our and Firmin's intellectual practice.

Carolyn Fluehr-Lobban (2006) and Greg Beckett (2017) have argued that Firmin is relevant to our contemporary struggles against racism. Politically, Fluehr-Lobban argues, Firmin is relevant for the "... systematic demystification of race, exposing the historic role that has played in the social dynamics of difference between more powerful and less powerful humans" (2002, xlii). Beckett suggests that we take care in embracing Firmin:

> ... not as a critic of race, but as a critic of racism. That, after all, is the promise and the challenge of his work. To embrace Firmin today, to invite him into our own intellectual genealogy after a century of exclusion, is to also push ourselves beyond the now comfortable sense that race is a social construction or a fiction (or just bad science) and into a more direct encounter with racism as a deeply embedded set of beliefs and practices.
>
> (2017, 174)

Beckett is careful to discard some of Firmin's views for two reasons: Firmin remained tied to the idea of biological race and remained committed to the Enlightenment idea of progress, as did other scholars of the age. Nevertheless, Beckett recognizes that in Firmin's criticism of how racist thinking led to erroneous anthropological hierarchies of races, we can also recognize how racism is embedded in our own concepts and categories (2017, 175). As two anthropologists who study race and racism and have attended multiple anthropology conferences and shared with a wide variety of colleagues, we believe that Fluehr-Lobban and Beckett's reclamation of Firmin should resonate with how most anthropologists today think about the potential of anthropological knowledge to be anti-racist.

Jafari Sinclaire Allen and Ryan Cecil Jobson reclaim Firmin as someone who provided a "foundational maneuver of a decolonial anthropology" by adhering to racial taxonomies "while denouncing the very existence of race as a biological type" (2016, 132). Allen and Jobson draw a dichotomy between Firmin and W. E. B. Du Bois, on the one hand, and Franz Boas and his students on the other hand. They argue that the former, by adhering to what we now perceive as an essentialist view of race were actually better positioned to undermine racism by being more attentive to the complexity of Black experiences. This is how Firmin is able to utilize his lived experience in Haiti as evidence for the equality of races. Meanwhile, Allen and Jobson argue Boas and his followers utilized the culture concept in such a way that rendered a weak criticism of racism as a social problem without fully attending to how Black folks lived meaningful lives through their racial experience. Following Visweswaran (1998), Allen and Jobson argue that cultural relativism left the notion of race intact as unchangeable and decontextualized it from how racism emerged as a tool for dominance and oppression in the Americas. It is important to point out that Allen and Jobson really emphasize the importance of "native" scholars, working within the field of anthropology, as essential to destabilizing anthropological worldview. Indeed, they locate the decolonization of anthropology within the discipline by anthropologists and scholars of color. This is why they can elevate Firmin, as a Haitian scholar working within a French anthropological institution in the early 1880s, as foundational to decolonizing anthropology.

Although we think these are valuable assessments and claims of Firmin for a contemporary anthropology, we have some doubts and criticisms about the manner in which his scholarship is positioned against contemporary anthropological theory, especially biocultural approaches. Fluehr-Lobban and Beckett want us to think of Firmin as someone through whom we can question power and racism in science and society. However, we do not need Firmin to question power and racism since we have plenty of more elaborate and complex social theories that encourage thought about power and racism. It is unclear in Fluehr-Lobban and Beckett's arguments how Firmin's work stands in relation to the many anthropologists and scholars in related fields who have theorized power, racism, and scientific knowledge. At best, they are holding Firmin as one of the earliest examples of scholarship that many social scientists carry out, especially scholars of color (e.g., McClaurin 2001; Collins 2008). Nonetheless, it would amount to nothing short of intellectual negligence to not acknowledge Fluehr-Lobban's role in introducing Firmin to anthropologists today. Her reading of Firmin against E. B. Tylor (2002, xxviii–xxxi), for example, is exactly the sort of intellectual historical method of inquiry we are after in this chapter; and although we seek to read Firmin against contemporary anthropological approaches to race, we view our work in this chapter as an extension of her initial effort.

Allen and Jobson more directly link Firmin to the work of contemporary anthropologists working on race. They coincide with Beckett in claiming that Firmin is an example of how scientific ways of knowledge can be blinded by societal racism, but Allen and Jobson link Firmin to historical and contemporary descendants and emphasize Firmin's embodiment of blackness as integral to Firmin's anti-racist anthropology. They pair Firmin with Du Bois and against the Boasian approach to anti-racist scholarship. We believe Allen and Jobson's interpretation of Firmin against Boas is incorrect. In our assessment of Allen and Jobson below, we point out some flaws in their argument and provide an alternative reading of Firmin because we believe that, in doing so, we can make a broader argument about the state and direction of anthropological theories about ancestry and race (and biocultural theory more generally). In other words, our criticism is meant to be a constructive engagement of Allen and Jobson and of many anthropologists who we think either subscribe or would subscribe to their views.

Allen and Jobson categorically claim that Firmin denounced "the very existence of race as a biological type" (2016, 132), but that interpretation projects a contemporary biological–cultural dichotomy into time period where it did not exist. By the time Firmin rejects the scientific traits used to distinguish human races (end of Chapter 5 in *Equality*), he is not prepared to reject the biological notion of race. Instead, he asks,

> ... we wonder whether the discipline is any more capable of providing a solution to a much more complex and difficult issue. Can one invoke anthropology to assert dogmatically that some of these human groups are congenitally and irremediably inferior to others? Is it possible to determine specific qualities before determining the species?
>
> (2000, 134)

After this point in *Equality*, Firmin goes on to demonstrate how environment can affect various aspects of the racial groups and how Egyptian history and contemporary métissage ("mixing") in Haiti demonstrate the ability of the Black race to be on par with that of Europeans. Accepting or not accepting the biological concept of race is not an option that Firmin entertains. Firmin works through the biological concept of race to argue for the equality of races and a world where the Black race can progress to high levels of civilization.

Firmin was careful in his scientific approach. Even when he is most frustrated by how anthropologists of the Paris Society abused of biological data, his analysis undoes bad science with what he thinks is better science. Allen and Jobson's claim that Firmin used the concept of race as a "Trojan horse"—to undue to concept of race by first embracing it—is not supported by Firmin's serious engagement with anthropological and evolutionary approaches to race throughout the book. Firmin redefines race as a different articulation of a species that, regardless of evolutionary origins, is equally capable of achieving high levels of civilization. Only environmental conditions, history, and European oppression can account for the differences in achievement of civilization between Whites and Blacks. Furthermore, we know that Firmin is serious about his engagement with race as a biological subject because he premises *Equality* with a resounding engagement of positivist science of humanity, which took racial groups as valid units of analysis. We might question Firmin's sincerity when he praises the scientific importance of the scientists whose racist work he criticizes, but Firmin sustained effort throughout his work to appeal to science probably means that he did admire scientific research as method of intellectual inquiry as much as he believed in the politics of liberation and anti-racism. He showed no more desire to do away with the concept of race than the concept of evolution—both indispensable concepts in the anthropology of the time.

In what we view as another presentist interpretation of intellectual history, Allen and Jobson differentiate Firmin from Boasian school and align Firmin with Du Bois and Frederick Douglass. However, there is more similarity in the structure of argument between Firmin's *Equality* and Boas's *Mind of Primitive Man* than there is between Firmin, Douglass, and Du Bois's published works. Firmin anticipates Boas's argument with a serious engagement of biological research and then utilizing historical and cultural examples to substantiate the equal worth of the Black race. Allen and Jobson's reliance on Visweswaran's reading of Boas as someone who inadvertently strengthened the race concept through the culture concept is unfortunate (see Lewis 2014 for a forceful defense of Boas against such contemporary criticisms). The main problem with this position is that it ambiguously establishes a causal relationship between the emergence of the culture concept and how people think about race—without intellectual historical empirical evidence and without taking into account the many other ways that race has persisted in societies. It is also an analysis that lacks cross-cultural validity since there is no explanation for how intellectual concepts of race existed in social and intellectual contexts (especially throughout the Americas) where Boasian anthropology had no measurable impact.

Like Firmin, Boas does not completely denounce race as a biological concept in his early scholarship because it was not scientifically possible to do so at the time (Torres Colón and Hobbs 2016, 127–29). Like Firmin, Boas limits himself to scientifically demonstrating that races cannot be ranked ordered. Moreover, it was Boas's students and other twentieth-century anthropologists who most accurately dismantled biological notions of race and racist biology. We agree with Allen and Jobson that Boasian anti-racist anthropology did not encourage a serious engagement with how people experience race (Baker 2010), but this does not mean that Du Bois and Firmin somehow were more effective than Boasian anthropology in rejecting biological notions of race. In fact, Du Bois's reliance on Boas's scholarship further complicates our intellectual historical analysis of the relationship between anthropological science and anti-racism. Du Bois draws on Boas's scholarship—as an anthropologist with solid grounding in evolutionary biology—in order to undermine scientific racism and provide a scientific backing to Du Bois's anti-racist politics.

Of course, there are very similar calls for Black liberation and equality in the works of Firmin, Du Bois, and Douglass; but Firmin's *Equality* is mostly filled with the same anthropological approach undertaken by Boas. Moreover, Firmin's approach to anthropology beyond biological/taxonomic comparison has the same intellectual roots as Boas in the figure of Alexander von Humbold (whom Boas's mentor, Adolf Bastian, followed in his naturalist and empiricist approach). Firmin's efforts for Black liberation beyond *Equality* are informed by the complexity of his experiences as a black Haitian, but his anthropology is empirically grounded just as Boas's critique of social evolutionism was empirically grounded and in line with the science of evolutionary biology (Herbert Lewis 2001). Allen and Jobson claim that Firmin and Du Bois "are accused of upholding a crude essentialism in contrast with the ardent deconstructionism of Boas" (2016, 132), but Allen and Jobson counter that it was Boas who was, in fact, unable to effectively critique the race concept—something that Firmin and Du Bois were able to do because "crude essentialism" is attentive to racial experience. But again, this is a misleading argument since Du Bois, by his own words and to some extent, was inspired by Boas's work:

> Franz Boas came to Atlanta University where I was teaching history in 1906 and said to a graduating class: You need not be ashamed of your African past; and then he recounted the history of the black kingdoms south of the Sahara for a thousand years. I was too astonished to speak.
>
> (2014, xxxi)

More importantly, Boas's critique of scientific racism was a necessity at a historical time when scientific racism directly influenced and had been used to legally oppress non-Whites as well as socially and economically marginalized Whites through eugenics. Yet the importance of anthropology as science went beyond simple critique of scientific racism, Du Bois continues in the same passage: "All of this I had never heard and I came then and afterwards to realize how the silence and neglect of science can let truth utterly disappear or even be unconsciously distorted" (2014, xxxi).

We believe that the leveraging of anthropology as an evolutionary, naturalist, and progressive science was as essential to Firmin as it was for Du Bois in a broader cultural effort of Black liberation. We are drawn to Allen and Jobson's sixth footnote, where they explain how biological anthropologist Michael Blakey (1998) is following Firmin when Blakey examines the relationship between concepts of race and racist sociocultural contexts. Blakey, however, does more than call our attention to how science is influenced by cultural notions of race and human diversity. Blakey calls for a "humanistic science" against "excessively naturalistic explanations of human society." This means practicing scientific theory that is critical of its own history, de-emphasizes the immutability of truth, seeks cultural and institutional diverse perspectives, acknowledges the sociopolitical vantage point of any approach, and reconciles any approach with one's accepted cultural values (1998, 386–87). Indeed, Blakey's career has been a model of a humanistic biological anthropology. In various projects (e.g., Blakey 1988, 2010), he has sought to practice a biological anthropology that is attentive to sociocultural context in the academy and in the society where he carries out research. Yet by Blakey's own reflective account of his research, a critical component of undoing decontextualized science is doing contextualized science. He reads this contextualization of science against "Positivistic notions of objectivity, extending from the Enlightenment to the modern era [that] foster an attempt at scientific disengagement from society that merely amounts to denial" (Blakey 1998, 402).

We think that Blakey does more than follow in Firmin's footsteps. In calling for a more humanistic science, Blakey is, in fact, continuing the work of many late nineteenth-century and early twentieth-century progressive naturalists who already saw the connection between human affairs and the natural world (Torres Colón and Hobbs 2015). Blakey is right in pointing to the ways in which scientific disengagement from society leads to bad science, but we think the answer is a science that is better reformed and more aware of how human experience shapes scientific practices. We take Blakey's humanistic science to simply be better science. Firmin, after all, subtitled *Equality*, "positive anthropology" and not "anti-positive and socially contextualized anthropology." He simply made contextualization an instrument of science—in 1885—which is how we envision a biocultural approach to race and genetic ancestry.

In fact, to call socially decontextualized science "positivist" can, inadvertently, yield the ground to what is "normal science" and continue to marginalize racialized worldviews as alternative forms of science. Our position, then, takes Firmin as an example for contemporary scientific practice as "positivist," and we characterize the science that Firmin fought against "racist science" or "racist positivism." By labeling the "normal positivism" as "racist positivism" and Firmin's anti-racist science as simply "positivism," *we are better positioned to carry out our work as anthropologists from a diversity of positions that are inclusive of scientific practices.* Most importantly, this conceptual shift places anthropologists of color and anthropologists working on issues of race at the center of the discipline and *forces us to* question *and denounce anthropologists that do not address questions of race relative to anti-racist anthropology.* We think this is a fair demand of a discipline that was born and theoretically developed in a racialized world.

Yet what we are calling for here is more theoretically significant than re-labeling anti-racist or humanistic science as science. The discipline of anthropology continues to suffer from lack of communication across subfields, and the anthropology of race is particularly vulnerable. In our experiences within the subfields of cultural anthropology and biological anthropology working on research question about genetic ancestry and race, we have noticed alarming blind spots within and between subfields. That Allen and Jobson fail to include any contemporary biological anthropological (or archaeological or linguistic) research in their decolonizing framework or to envision a role for it is not an indictment on their scholarship but rather reflective of the status quo of intellectual divisions of labor. The problem is even more acute within the subfield of cultural anthropology, where humanistic and scientific cultural anthropologists continue to ignore each other, as evidenced by a lack of citations and theoretical engagement. Therefore, in claiming Firmin as an ancestor, we claim his anti-racist fervor, his liberatory politics, and his integrative scientific anthropology. Let us now examine how Firmin's legacy resonates with the contemporary experiences of biological anthropologists from racialized communities.

Science, progress, and anti-racism

The promise of science as a pathway to knowledge that could contribute to improvements in the lives of Black people was commonly shared by Firmin, Boas, and Du Bois. Where we see stark differences between these three intellectual figures is in how they leverage the knowledge to achieve equality among humans. For Boas and his followers' knowledge and appreciation of human cultural difference was fundamental for a political project of equality. For Firmin and Du Bois, knowledge of human equality was but one step in more complex political efforts that ranged from educational opportunity to liberation. However, Firmin and Du Bois, along with many other intellectuals of their time, all believed in the progress of Black people. This notion of "progress" is often muddled in presentist interpretations of intellectual history, for is it not racist to assume, as Firmin did, that many Black people had yet to achieve high levels of civilization? Anthropology, according to Firmin, was instrumental in establishing a fundamental baseline of equality from where to argue for political positions aimed at achieving that equality. However, the development of anthropology as a discipline in the twentieth century complicated the potential of racial improvement by pursuing intellectual projects that, through the concept of cultural relativism, sought to establish the equal worth of all cultures. If all cultures are equal, then why would we seek to change them or improve them?

The answer, as Lee Baker points out, was often found in the dismissal of many peoples of African descent as not having a culture (Baker 2010). Within the field of anthropology, we can think of this dilemma as a dissonance between local and global diversity (Torres Colón and Hobbs 2016). Local racial diversity was characterized by Black and other racialized peoples whose condition needed improvement (i.e., civilization), while global diversity (even when seen through the lens of "local" Native Americans in the United States) could be understood, appreciated, and preserved. The problem with early anthropology is not that it shifted negative ideas of racial groups from the biological to the cultural, the problem is that either marginalized racialized groups were deemed improvable or they were appreciated for their cultural worth in their marginalized condition. The disciplinary answer to this dilemma is that race—especially Black people—were taken up by sociologists as a social problem to be solved and culture remained the domain of anthropology. This historical occurrence, which we take for granted, should be cause for alarm for all anthropologists. The most culturally immediate social category for recognizing social difference in the United States—race—was bracketed as theoretically undesirable and anthropology developed through a concept—culture—that existentially excluded the Black people that lived in the same cities where American anthropology took shape.

We think this intellectual historical understanding is the best framework for understanding why Du Bois concluded that "the silence and neglect of science can let truth utterly disappear or even be unconsciously distorted" and Firmin reasoned that businessmen in Haiti were instrumental for the progress of the Black race. Speaking about business skills In *Equality*, Firmin insists

> ... that the qualities I am praising here are of the utmost importance in any inventory of ethnic aptitudes. Black men must be convinced of this one fact: they will be recognized as equals by all other men in terms of their abilities only when they have achieved material success and accumulated wealth, while at the same time achieving intellectual success and accumulating knowledge.
>
> (2000, 223)

Du Bois and Firmin were successful at articulating and fighting for an anti-racist politics because they embodied the level of civilization that they desired for all Black people. This is also why Du Bois can question the problematization of Blackness while also thinking about how to address the problems faced by Black people. That level of civilization inevitably linked the economic improvement of communities with improvements in knowledge. How these improvements were prioritized was the subject of much debate among intellectuals through the African diaspora, but the improvement was nonetheless desired. Science, as an instrument of knowledge, was integral to this agenda; and the ability of Du Bois and Firmin to recognize science as a tool of progress should lead us to more complex assessments of the role of science in perpetuating racism in the past. If science was leveraged in the name of racial hierarchies, then better science was leveraged in the development of anti-racist anthropology and anti-racist politics more generally.

Biological anthropology and theories of race and anti-racism

If, as Allen and Jobson observe, "The possibility of an anthropology for liberation requires that the discipline attend equally to its conceptual registers and professional codes in its advance toward the ideal of decolonization" (2016), then it is imperative to include anthropological perspectives that, following Firmin, utilize science in the service of decolonizing anthropological theory. Doing away with science as a Western form of knowledge opposed to non-Western ontologies or opposed to the decolonizing work of anthropologist of color in non-scientific disciplines is simply not acceptable because it is an argument with flaws in historical analysis and inattentive to theoretical developments in anthropology, particularly integrative approaches to bio-cultural research. As we have shown, Firmin lays the foundation for anti-racist anthropology before anthropology was institutionalized as a four-field discipline, and he wielded a positivist approach before dominant anthropological theories and professional practices excluded perspectives from racialized peoples. Therefore, a scientific anthropology is not the elephant in the room, but people of color theorizing and doing anthropology are the elephant in the room. In what follows, we reflect on our own experiences and conversation with scholars of color working in biological anthropology and related biomedical fields to rearticulate the role of science in anti-racist scholarship.

"I am not a scientist. I am a Black scientist." Those were the exact words of a top Black biomedical scientist working on racial health disparities. These words also echo the sentiments of others with whom we talked about how science and anthropology are intrinsically linked to race. More than ethnographic work, when we reached out to colleagues for interviews, our conversations felt more like continued conversations that we have all had since we entered graduate school. These conversations often take place away from common professional settings like conferences, departmental meetings, and classrooms. Even the occasional diversity forums at professional conferences do not yield the stark level of contrast between anthropology and an anthropology of color that we have noted in private conversations. Those forums are meant to bridge diverse perspectives and diversify anthropological thought. The notion of a Black scientist, however, is much more personal and meant to establish a different reason for doing science, for doing anthropology. In fact, the notion that science or anthropology are common practice through which we can achieve different goals is thrown into question. Instead, the very practice of science and anthropology are understood as Black and anti-racist—just as Firmin's anthropology was articulated from a Black Haitian epistemology. There are several dimensions to this perspective.

Biological anthropologists from racialized communities that we spoke with think that the notion of scholarship—research and teaching—as a means to combating racism is something they have in common with anthropologists and other scholars outside of biological anthropology. As scholars of color, they share a general passion for bringing diverse perspectives into anthropology. With all anthropologists, they share a more general passion for understanding humanity from cross-cultural and holistic perspectives. However, they have consistently faced scholarly critiques of racism that highlight the role of science in promulgating racist worldviews. In formal and informal social situations with colleagues, they often walk away feeling that they occupy a compromised scientific world that should be scrutinized for various research tropes that support racist views. These tropes include genetic determinism, ad hoc evolutionary explanations, socially and culturally decontextualized explanations for biological differences, and classifications of human diversity. Even though the biological anthropologists from racialized communities that we spoke with do not hold these views, they often feel like their peers in other subfields believe that they support such views and are on "the wrong side" of scholarship. And to add insult to injury, they have explicitly heard and implicitly perceived that they, as racialized people in science, are ignorant of the role of science in racist worldview and have blindly entered a discipline that is used as a tool of oppression.

Simply and unequivocally stated, in our interviews, biological anthropologists from racialized communities do not believe that science is inherently racist. They are the ones who have variously navigated National Institutes of Health (NIH) granting processes to carry out research on race, even when the NIH can both reinforce the uncritical use of race in research and deny efforts to undermine alternative biomedical approaches to race. They are also the ones who have managed to hold anti-racist politics and carry out scientific work. This latter point emerged as a problem for some of the scientists who we talked to because they felt that doing biological work in science was so difficult that often people from racialized communities end up working in areas (such as bioethics) that are significantly different from the training they received as graduate students. They also resent colleagues who talk about the necessity to do science and not reinforce racist worldviews without actually ever having to do research in which race has to be operationalized in scientifically valid ways. In other words, it is far easier to say there is no biological basis to race than engage with scientific topics that involve race. This has created feelings of despair that are amplified when reading how colleagues from racialized communities outside the biological sciences craft scholarly visions for a more diversified discipline without clearly articulating the role of a biological anthropology theoretically shaped by racial experience in such a vision.

These experiences with non-biology colleagues have led to many instances of personal and professional frustration where the biological anthropologists we spoke with, who have received extensive training as biologists, are made to feel as guests in conversation about race and biology. In response to marginalization within academia that scholars from racialized communities experience, some senior researchers have purposefully created academic environments in which Black, Latino, and other minority students are welcomed to learn and practice science. These are actual laboratories where racialized minorities are significantly overrepresented relative to other biomedical laboratories. In these spaces, science is carried out with awareness for external critiques, but also with a resigned determinism of organically training new generations of biomedical researchers and formulating an anti-racist biology from within scientific spaces. It is important to note, that even when biological anthropologists are heading or play a significant role in these laboratories, there is skepticism about the possibility of future work as biological anthropologists—hence the necessity to target future professional work in "biomedicine." Yet the most important element of these laboratories is the formulation of anti-racist science from within science. These formulations are destined for the communities with whom Black biological anthropologists work, and there is less concern about how they might resonate with colleagues, anthropology, and science in general. Whether these anti-racist scientific formulations end up as interpersonal conversations with research participants, community forums, or local exhibits, the limited reach of science that makes a difference in the lives of racialized peoples is more important that holding the limelight as exemplary anti-racist academics.

Another interesting aspect of doing anti-racist biological anthropology is related to the intense academic debates over modern populations and racial terminologies. For example, Black biological anthropologists often work with racialized populations. The scientist who identified herself as a "Black scientist" rationalizes her identity through the concern and work she does with racialized peoples. Moreover, all of the people who we interviewed have lifetimes of experiencing the world as racialized minorities—in and out the academy. As an extension of that experience, some have even articulated clear distinctions between racial terms and more appropriate ways of labeling genetic populations (Benn Torres and Kittles 2007). Yet outside of biomedical research, the anthropological community has essentially ignored this work; and the alternative terminology for describing diversity in human populations—such as "biogeographical ancestry"—is more likely to be targeted by philosophers and social scientist of science (e.g., Frank 2007; Gannett 2014) than adopted and dissimilated by non-biological anthropologists writing about race. Although there is little disagreement about the inadequacy of biological races, concepts that have been adopted by Black scientists as alternatives to race seem to have little coinage beyond their already small scientific worlds.

Another Black scientist questioned their lack of influence on conversations about race by wondering who, if it was not Black scientists, were going to find alternative ways of talking about the biology of "Black people." In many of our conversations, Black scientists and biological anthropologists from racialized communities used racial terms, like "Black" (along with "Brown," and "White," and Native American) in order to describe the very same populations that they argue cannot be biologically classified with racial terms. Clearly, they are employing a sociocultural understanding of race in these instances, but for them, this is more than an internalized distinction, "I have to leave the lab and still be Black in the world." Contemporary anthropological rhetoric of anti-racism—whether coming from White, Black, or Brown anthropologists—simply does not resonate with Black scientists' lived experiences as Black people and Black scientists. For the experiences of being a Black, Latinx, or Native scientist adds an eerie dimension to how people experience race: race is not biological, yet they study racialized people from biological perspectives that are themselves racialized. To put it differently, once race is undone (deconstructed) from traditional and contemporary perspective, scientists from racialized communities are still left with having to think about their lived intersection of biology and race; and pedagogical anti-racist victories of students and general public are simply not as satisfying.

At this point, we can return to Firmin and listen to his voice as a Black Haitian anthropologist who simultaneously embraced scientific, legal, and political reasoning in the struggle for Black liberation. There is no need to annotate his legacy with warnings about his subscriptions to positivism and the progressive promise of the French Revolution. His science cannot be discarded. In the next chapter and along with our Caribbean interlocutors, we articulate a contemporary version of Firmin's legacy with regard to genetic ancestry research and anti-racism.

References

Allen, Jafari Sinclaire, and Ryan Cecil Jobson. 2016. "The Decolonizing Generation: (Race and) Theory in Anthropology since the Eighties." *Current Anthropology* 57 (2): 129–48.

Baker, Lee D. 2010. *Anthropology and the Racial Politics of Culture*. Durham: Duke University Press.

Beckett, Greg. 2017. "The Abolition of All Privilege: Race, Equality, and Freedom in the Work of Anténor Firmin." *Critique of Anthropology* 37 (2): 160–78. doi:10.1177/0308275X17694945.

Benn Torres, Jada, and Rick A. Kittles. 2007. "The Relationship between 'Race' and Genetics in Biomedical Research." *Current Hypertension Reports* 9 (3): 196–201. doi:10.1007/s11906-007-0035-1.

Blakey, Michael L. 1988. "Social Policy, Economics, and Demographic Change in Nanticoke-Moor Ethnohistory." *American Journal of Physical Anthropology* 75 (4): 493–502. doi:10.1002/ajpa.1330750407.

———. 1998. "Beyond European Enlightenment: Toward a Critical and Humanistic Human Biology." In *Building a New Biocultural Synthesis: Political-Economic Perspectives on Human Biology*. Alan H. Goodman and Thomas L. Leatherman, eds. 379–405. Ann Arbor: The University of Michigan Press.

———. 2010. "African Burial Ground Project: Paradigm for Cooperation?" *Museum International* 62 (1–2): 61–8. doi:10.1111/j.1468-0033.2010.01716.x.

Boas, Franz. 1911. *The Mind of Primitive Man: A Course of Lectures Delivered Before the Lowell Institute, Boston, Mass., and the National University of Mexico, 1910–1911*. New York: The Macmillan Company.

Charles, Asselin. 2014. "Race and Geopolitics in the Work of Anténor Firmin." *The Journal of Pan African Studies (Online)* 7 (2): 68.

Collins, Patricia Hill. 2008. *Black Feminist Thought: Knowledge, Consciousness, and the Politics of Empowerment*. New York: Routledge.

Du Bois, W. E. B. 2014. *Black Folk Then and Now (The Oxford W.E.B. Du Bois): An Essay in the History and Sociology of the Negro Race*. Oxford: Oxford University Press.

Firmin, Joseph-Anténor. 2000 [1885]. *The Equality of the Human Races: (Positivist Anthropology)*. New York: Garland Pub.

Flourens, M. 1861. *Eloge Historique de Tiedemann*. Paris.

Fluehr-Lobban, Carolyn. 2002. "Introduction." In *The Equality of the Human Races*, edited by Anténor Firmin, xi–xlvi. Urbana: University of Illinois Press.

———. 2006. "Anténor Firmin: His Legacy and Continuing Relevance." In *Reinterpreting the Haitian Revolution and Its Cultural Aftershocks*, edited by Martin Munro, and Elizabeth Walcott-Hackshaw, 86–101. Kingston: University of the West Indies Press.

Frank, Reanne. 2007. "What to Make of It? The (Re)Emergence of a Biological Conceptualization of Race in Health Disparities Research." *Social Science & Medicine* 64 (10): 1977–983. doi:10.1016/j.socscimed.2007.01.010.

Gannett, Lisa. 2014. "Biogeographical Ancestry and Race." *Studies in History and Philosophy of Science Part C: Studies in History and Philosophy of Biological and Biomedical Sciences* 47 (September): 173–84. doi:10.1016/j.shpsc.2014.05.017.

Lewis, Herbert. 2001. "Boas, Darwin, Science, and Anthropology." *Current Anthropology* 42 (3): 381–406. doi:10.1086/320474.

———. 2014. *In Defense of Anthropology: An Investigation of the Critique of Anthropology*. New Brunswick: Transaction Publishers.

McClaurin, Irma. 2001. *Black Feminist Anthropology Theory, Politics, Praxis, and Poetics*. Black Women Writers Series. New Brunswick, NJ: Rutgers University Press.

Marks, Jonathan. 2009. *Why I Am Not a Scientist*. Berkley: University of California Press.

Torres Colón, Gabriel Alejandro, and Charles A. Hobbs. 2015. "The Intertwining of Culture and Nature: Franz Boas, John Dewey, and Deweyan Strands of American Anthropology." *Journal of the History of Ideas* 76 (1): 139–62.

———. 2016. "Toward a Pragmatist Anthropology of Race." *The Pluralist* 11 (1): 126–35. doi:10.5406/pluralist.11.1.0126.

Visweswaran, Kamala. 1998. "Race and the Culture of Anthropology." *American Anthropologist* 100 (1): 70–83. doi:10.1525/aa.1998.100.1.70.

6 Genetic ancestry as empowerment

Scholars have levied a number of serious and substantial critiques against genetic ancestry testing. Notably, it has been implicated in the reification of biological race concepts, of undermining community sovereignty by resetting the terms of how a community might identify its members, as well as the many subjective aspects that go into designing and interpreting an ancestry test (Krimsky and Sloan 2011). These critiques are certainly valid and warrant careful consideration with the use of ancestry technologies. However, we think it is also important to be cognizant of the many different contexts in which genetic ancestry tests might be deployed and the utility of these types of tests with regard to lived experience. In our research, we have found evidence that the mentioned critiques are indeed credible; however, we have also found that genetic ancestry can be utilized as a tool for social and political empowerment. The two previous chapters are most critical for beginning our discussion of this concept for theories of race as bioculturally embodied difference and the ideas of scientists from racialized communities are essential components of genetic ancestry as empowerment. In our explanation of the utility of genetic ancestry as empowerment, we return to our fieldwork among the Accompong Town Maroons.

As discussed in Chapter 3, we spent the summer of 2011 in Accompong Town, in the western mountainous region of Jamaica. Since winning their independence from the British in 1739 (Patterson 1970), Accompong Maroons have an attachment to land that has harvested a sense of indigenous belonging in Jamaica, the Caribbean, and the Americas (Zips 1998; Thompson 2012). We call this attachment to land "displaced indigeneity" because it leads Maroons to naturalize their collective existence as autochthonous to their land—a subtle yet significant form of embodied difference that distinguishes them from a surrounding population of Jamaicans with whom Maroons share an ancestral past (cf. Anderson 2009). We characterize their indigeneity as "displaced," not because we locate their proper place of indigeneity in Africa, but because they themselves locate Africa and their community's origins on the island in their origin stories. Rather than projecting an exotic image of the folkloric Other, Accompong Maroons report experiencing bodily Otherness when other Jamaicans characterize them as devils, or as dangerous, backward, tough, and cagey.

When we approached the people of Accompong Town with a genetic ancestry research project, we had three general goals in mind: (i) engage and incorporate community perspectives into a genetic ancestry project, (ii) study the genetic ancestry of the community, and (iii) frame their genetic ancestry within the historical reality of maroonage. With the permission of community leadership and in conjunction with local research assistants, Jada collected 50 DNA samples from consenting community members and ascertained ancestry using mitochondrial, Y chromosome, and autosomal markers (Madrilejo, Lombard, and Torres 2014; Fuller and Benn Torres 2018). The maternal, paternal, and general genetic lineages provided insights into the biogeographic ancestry of Accompong Town Maroons as well as illustrated how genetic ancestry can be wielded in ways that are responsive to community-engaged research. Based on these studies, we concluded that Accompong Town Maroons, as expected, have primarily African ancestry and, to a lesser extent, some European and Indigenous American ancestry. The findings of our genetic analysis aligned with community oral histories regarding the origins of their community within Africa and the Americas and subsequent influx of European ancestry. Several community members explained that their African and Indigenous American ancestors were unified in resisting colonization and enslavement. European ancestry was explained as a consequence of the 1739 treaty in which Accompong Town Maroons had to concede to having the presence of English soldiers within the community or as a result of more recent relationships between Maroons and non-Maroons.

Participant reaction to the study findings varied, from excited to completely indifferent. Some participants were content to discuss their individual results seemingly reviewing them as a means to quell their curiosity about genetic ancestry testing. Yet other participants' curiosity cultivated discussions about the potential ramifications of the study leading some to speculate on what, if any, impact the genetic ancestry results might have on solidifying Maroon claims to the land. Other participants, as illustrated in the vignette about Selwyn in Chapter 3, simply expressed their thanks for returning results then quietly folded up and stashed the report among their belongings. These varied reactions to genetic ancestry reiterate our purpose in doing ancestry work: we did not go to Accompong Town to rewrite Maroons's sociopolitical existence from a scholarly perspective. Instead, guided by our participants, we allowed our research and writings to be framed by the spirit of resistance and liberation that was already there, albeit in complicated ways that escape simplistic romanticization. Our research was co-opted by the community as much as we co-opted notions of "maroonage" to frame our genetic ancestry research. Our experience in Accompong Town, among other places, illustrates a broader point we intend to convey, that despite the very real interpretative limitations, genetic ancestry testing ultimately is a tool that can be differently valued or discarded. In the remainder of this chapter, we extrapolate on this point relying on the theoretical approaches of our forebears such as Firmin to show the potential in ancestry testing can impact the manner in which we think about our ancestors and our past from local and scientific perspectives.

Resistance: defining a population

Genetic ancestry as empowerment is not meant to serve as a metaphor. Like Firmin writing from a scientific standpoint against scientific racism, we propose that genetic ancestry as empowerment represents a more scientifically sound approach to doing genetic ancestry research. As such, we begin by taking aim at the concept of population, which is central to the evolutionary and cultural study of humanity. What constitutes a population is also a critical starting point for doing genetic ancestry research for both setting the parameters of the studied population and for the referent populations used to estimate ancestry (see Chapter 1). Surprisingly, how one defines a population is not as easy as opening up an anthropology textbook and looking up a definition or a methodological guide.

In his "Introduction to Molecular Anthropology," Mark Stoneking discusses some basic characteristics of a human population (Stoneking 2017). Stoneking notes that a population must have a spatiotemporal social dimension and that members of the population must make offspring together. The spatiotemporal social dimension simply indicates that a population has existed for several generations in a shared geographical space. Within this geographical space and over generations, individuals will have had access to one another as mates. Determining what counts as a biological human population, then, requires close attention to social and historical conditions, which means that anthropological fieldwork becomes of utmost importance to appropriately delineate specific population parameters. Crawford (2007) argues that fieldwork is necessary to more accurately assess gene–environment interactions, ascertain historical time dimensions for interpreting genetic data, and possibly uncover the etiology and mechanisms of genetic transmissions involving complex phenotypes. Similarly, he argues, sampling strategies should be attentive to socio-demographic factors discovered during fieldwork, especially if random sampling of a population is not possible.

We can assess these fundamental approaches to defining a human population and sampling strategies through the biocultural prism we discussed in Chapter 4. Rather than seeing human populations as either purely social phenomena or purely biological entities that naturally manifest themselves as races or nations, we can recognize that social and historical processes configure human mating patterns. Generations later, these mating patterns can be observed genetically because individuals tend to find mates that are geographically near to them and this results in an overall pattern in which individuals within a population tend to share alleles with each other at a higher frequency relative to individuals outside of that population and who are geographically distant. At this point, it might appear that we are contradicting the idea that race has biological foundations, which we are not—but this is where a faithful application of biocultural models starts getting complicated with regard to race.

Recall from Chapter 1 that we can examine genetic ancestry through the use of ancestry informative markers (AIMs) and by relying on broad patterns of population substructure. The ability to use genetic data to identify ancestry, however, is hardly indicative of biological racial groups given the clinal nature of genetic variation and the over-all genetic similarity between members of our species. Rather, the fact that one population may have particular alleles that occur at higher frequencies than another population is a biological fact that is reflective of evolutionary processes such as genetic drift, natural selection, as well as factors that have shaped how people picked mates over many generations. Though it can be used to assess ancestry, this difference in allele frequencies among AIMs is not supportive of a biological ba-sis to race, but rather is the consequence of sampling from distinct regions along a cline (Chapter 3). To reiterate, race is a product of culture, and though it may be intimately tied to biology, race is not biology. However, the resonance between race and biology is biocul-tural in nature—and therefore—it is supportive of the idea that there are biological aspects of race that anthropologists have obfuscated by theoretically ascribing them as solely sociocultural and wholly separate from biology. As we have argued elsewhere, in attempts to reconcile the racist science that characterized early biological anthropology, contemporary anthropological perspectives on race have not adequately addressed how to theoretically bridge biological variation with racial experience (Benn Torres and Torres Colón 2015). We can both affirm that human biological diversity is not causative of culturally conceived racial groups, yet we can also acknowledge that culturally conceived racial groups can correspond with subsets of human biological variation.

Now, even if we accept a view of human biological variation that allows us to think about how the social, cultural, and historical dimension of race can shape genetic variation within a given social geographical space, we still have to deal with the task of defining and delimiting that social geographical space, that is, a population, before collecting samples. Sociocultural anthropologists have for a long time understood that, depending on social and (pre)historical context, there are a myriad of ways to think about populations. Some of these ways are related to the different forms of ancestry that we explored in Chapter 2. Kinship groups, nations, and groups marked by embodied difference often exist in delimited geographical spaces over long periods of time. However, there are many ways in which humans organize themselves into groups that we might want to call "populations." For example, kinship groups alone present dozens of possibilities that vary throughout human history and around the globe. However, we can make some basic observations about how to draw parameters for thinking about populations. Here, as we did in Chapter 2 for thinking about relatedness, we can think about populations from social and symbolic dimensions. Social dimensions include economic systems, political organizations, kinship systems, and groups of embodied difference. Symbolic (or cultural) dimensions often overlap with these social dimensions since humans are aware, more or less, of their social organizations and give these systems symbolic value. These symbolic dimensions include communities of people who speak the same language, groups with a common history and heritage (e.g., ethnic and national groups), religious groups, and political groups.

Some anthropologists (e.g., Sahlins 2000) claim that social systems are always symbolic and that the concept of culture encapsulates both the social and the symbolic. We agree with this theoretical position; but here we want to make an initial distinction between the social and symbolic because, if we want to define a population for genetic ancestry research, then we want to be aware of the relationship between social systems and patterns of gene exchange. Of course, we also want to be aware of the symbolic value that humans give or do not give such social systems and ask questions about how such valuation has anything to do with the degree of gene exchange (i.e., mating patterns and norms). For example, it is quite common for human groups to support one another economically without exchanging mates. Yet economic exchange is integral to subsistence, so the biological continuity of a population can depend on another population that is not contributing genes. Therefore, if two hypothetical populations do not exchange mates, we can still see them as having biological ties if their survival is dependent on economic exchange. At face value, such distinctions would not matter from the perspective of population genetics, since only those groups that exchange genes, that is, are breeding populations, matter with regard to understanding the dynamics of genetic change within and between populations. However, understanding the complexities of how populations are defined and define themselves is an important component in any attempt to draw conclusions about genetic changes within and between populations.

Because populations are not simply a static group of people occupying a geographical space, people have their own sense of how populations work and came into existence. These are "folk theories" (Gelman and Legare 2011) about populations, and they are more than just a notion of "who we are," but also includes notions of history, ethos, and biology. In everyday speech, the word "population" itself is often symbolically unambiguous and attached to a rationalized group of people. In the Spanish language, the word population is "población," which is derived from the verb *poblar* (to populate) and *pueblo* (people, village, folk, or populace). When politicians and citizens of many Spanish-speaking countries speak of what "el pueblo" wants, they are making strong claims about the character of a population. In the English language "population" does not resonate with such fervor, yet the word "people" does carry symbolic weight. For example, "We the people" is enshrined in the US Constitution. Indeed, the words population, *pueblo*, and people all come from the same Latin root, *populus*, which can refer to a group of male citizens, a population, or a nation.

In stratified societies, there are always multiple senses of population since social relationships are differentiated by economic activity, political power, geography, and cultural variation (e.g., language, regional identities). Because economic power is often dependent on exploitation of natural and human resources, controlling peoples over vast geographical spaces is necessary; so those who have interest in control over those broad spaces impose their sense of population over socially diverse regions. Government and governing represent vested interests in defining who constitutes "the people" in order to have political legitimacy in acting politically on behalf of "the people." Educational systems, maps, and the census are some of the ways through which governments can obviate and naturalize the congruence between "the people" and the state (Inda 2005; Anderson 2006). These governmental ideas about who constitutes "the people" are no more valid than "folk theories" about populations. In fact, we can think of governmental definitions of population as folk theories of the powerful; the difference, however, is just that—theories of the powerful backed by institutionalized legitimacy and economic interests. Therefore, when genetic ancestry research is paired with alternative and community-based notions of population that contradict dominant ideas about populations, we can see how genetic ancestry research can serve as resistance.

The differences and similarities between social, symbolic, and folk theories of population are different ways of approaching human existence. These different approaches are not more or less true because from an anthropological perspective, they all represent an important aspect of people's sense of meaning in relation to other people. *Instead, whatever differences and similarities exist between social, symbolic, and folk population are empirically questionable dimensions of the complex processes through which humans live out their sociality.* For example, humans might have economic relationships that are not recognized through ethnic ties, have racial identities that do not recognize biological admixture, or have national identities despite massive economic rifts. When we define a population for genetic ancestry research, we have to be aware of and take into account the social and symbolic complexities that shape people's experiences or risk misrepresenting the reality of the study community. Genetic ancestry can shed light on particular aspects of population history, but these aspects of population history cannot be assumed to serve as a corrective to false senses of the past. Of course, there are many instances in which a group makes claims about their biological population history (e.g., claims to "racial purity"), and in those cases, genetic ancestry technology makes truth claims through a scientific process that is more universal and replicable—thus having better claims to an objective truth. However, note that social, symbolic, and folk theories of populations are also "objective truths" that can be empirically discovered by social scientists. Most importantly, genetic ancestry tests rely on humans to collect samples; and those humans need to have a working definition of population from where they can sample.

Our research with Afro-Puerto Rican populations demonstrates how defining a population for genetic ancestry research can take into account the complexities of different social, symbolic, and folk theories of populations. In circumscribing our study population this way, we can address questions that go beyond descriptive analyses of various genetic features and attempt to understand how social, historical, political, or economic contexts have shaped patterns of genetic variation within the population. This approach to anthropological genetic research and consequently genetic ancestry can be a form of resistance and empowerment.

Pues negro porque negro soy ("well Black, because I am Black")

In the early 2000s Dr. Juan Martínez-Cruzado and colleagues first published their reconstruction of Puerto Rico's population history through genetic ancestry analysis (Martínez Cruzado et al. 2001, 2005). Their work made headlines in Puerto Rico's major newspapers. Knowing many people from fieldwork and familiar ties, we can assess that many, if not most, formally educated Puerto Ricans have some knowledge of these studies; and they can have a meaningful conversation about how those genetic findings make sense within the context of Puerto Rican history. Both scholarly research and the press emphasized the surprising estimates that two-thirds of Puerto Ricans have Taíno (Indigenous Caribbean) mtDNA ancestry. There was some controversy regarding these studies at the scholarly level. Specifically, Estevez (2008) argued for the significance of DNA evidence in countering historical narratives of Taíno extinction, while Brusi-Gil de Lamadrid and Godreau (2007) and Haslip-Viera (2008) cautioned against the ways in which emphasis on Taíno identity has been historically exclusive of African ancestry and living black culture on the island. This latter point was echoed in informal conversations when we began to first visit Afro-Puerto Ricans around the island in the spring of 2017.

Our ongoing research project is primarily based in Piñones, which is politically part of the municipality of Loíza, a town in the northeastern region of Puerto Rico and immediately east of San Juan. People from Piñones and Loíza often distinguish themselves from one another—mainly as a result of their historical occupation of either side of the Río Grande de Loíza. Piñones lies west of the river and is mostly composed of a mangrove forest. Most of the people of Loíza and Piñones are black and descendants of both free and liberated and self-liberated African peoples. However, it is important to explain this latter point because saying that all black people throughout the Americas are descendants of enslaved peoples (and self-liberated or liberated enslaved people) is both inaccurate and incredibly vague. As the historian Juan Giusti Cordero (2015) has argued for Piñones, not recognizing the nuances of how people in Piñones are economically and ecologically related to their geographic space contributes to a long historiography of giving significance to Afro-Puerto Ricans only though folklore and slavery. His works not only cautions against seeing all Afro-Puerto Ricans as historically tied to sugarcane production, but he explains how intimate knowledge of the mangrove forest and sustainability can help us see the black people throughout the Americas as much more than their blackness as a result of the Transatlantic slave trade.

Our interest in working with Afro-Puerto Rican communities was twofold. On the one hand, genetic ancestry research could help deepen our historical understanding of peoples in the Caribbean and also lay the foundation for research in understanding the roles of race within the context of women's health and health disparities. On the other hand, engaging people through fieldwork about their history could serve as a vehicle for collaborative research about the nature of racial experience in Puerto Rico; furthermore, such research could inform policy for meaningful social change in a community that has been marginalized since colonial times. More important than our interests, we wanted to listen to community members and leaders to make sure that the project was theirs as much as it was ours.

Once in Piñones, we immediately found that ancestry research (historical and genetic) was conceptually congruent with Afro-Puerto Rican efforts to recover and uphold Afro-Puerto Rican traditions. Researchers and community leaders coincided on the sentiment that popular portrayals (i.e., those outside of Piñones) of Afro-Puerto Rican traditions emphasize the significance of Black music, cuisine, and poetry in ways that actually conscript Afro-Puerto Rican ancestry to folkloric caricatures (see Godreau 2015 for an excellent and in-depth study of Blackness in Puerto Rican popular culture and politics). However, in Piñones, these aesthetic cultural forms are tied to everyday struggles for equality and social justice. The musical tradition of "bomba," for example, is not simply a component of Puerto Rico's tricultural heritage (European, Taíno, and African). Instead, it is a way for Afro-Puerto Ricans to tell stories about themselves and assert their autonomy in response to everyday experiences of discrimination. Our genetic ancestry project would similarly contribute to popular historical explorations of "who we are" as Puerto Ricans while at the same time contributing to political assertions about the importance of African heritage in Piñones. Those political assertions, in turn, are tied to specific political action regarding racial discrimination, remedying poverty, and struggles to protect Piñones's environment from industrialist encroachment (e.g., big hotel companies and expansion of Puerto Rico's international airport).

Of course, not all Afro-Puerto Ricans in Piñones embrace traditional Afro-Puerto Rican aesthetic traditions (indeed, some reject it as antiquated), but people in Piñones do have a sense of place and community that makes them feel like their African heritage is theirs in a way that is more intimate and alive than it is for most Puerto Ricans who are not black. Again, as Giusti Cordero (2015) reminds us, labor and ecological knowledge have tied the people of Piñones to the land for centuries; and those intimate and local ways of knowing Piñones persist into modern ways in which the people of Piñones make a meaningful living. Whether people participate in Afro-Puerto Rican artistic traditions or not, many people participating in our project simply wanted to know about their past and further deepen their knowledge of those whose memories are tied to Piñones. Moreover, our ideas about future research on women's health and experiences of race that could inform political projects extended well beyond the history and politics of recognizing Blackness.

We assumed very little with regards to defining an Afro-Puerto Rican population. Months before we collected our first sample, we took to the field to talk to different people, from all walks of life, about their sense of community in relation to being black. Several potential problems in defining a population emerged immediately. First, what do we call the population when the terms for Blackness varied? The terms Afro-Puerto Rican or Afro-descendants are not universally embraced. Although less controversial for black Puerto Ricans, it is often rejected or seen with suspicion by other Puerto Ricans. In fact, Gabriel was "informed" on multiple occasions by other non-black Puerto Ricans that "Afro-Puerto Rican" is not an accurate or relevant term—as if it was us the anthropologists who were importing a derivative of "Afro-American" from the United States to Puerto Rico. This was awkward for Gabriel because he was, as a Puerto Rican living abroad, encountering (as an anthropologist) people in Puerto Rico who identified as "Afro-Puerto Rican," and then being told by other people on the island that such identifications did not exist. It is true that there are black Puerto Ricans who also reject the term "Afro-Puerto Rican," but they usually reject the term based on the grounds that it is politically charged with anti-racist or Afrocentric movements; yet they tend to accept the blackness of their skin color.

Furthermore, social and geographical spaces seemed to be difficult to locate outside of Loíza. We say "seemed" because after talking with enough people, we realized that Afro-Puerto Ricans do have a sense of where other Black Puerto Ricans live; and these places are usually specific to regions, towns, and neighborhoods.

We decided to tap into the social network of community leaders and artists (who are often one and the same) around the island and allow them to meaningfully shape how we conceived of Afro-Puerto Rican communities. Since every genetic sample we took was accompanied by extensive interviews and many also included life and community histories, we figured that if community and artistic leaders' conceptions of who is Afro-Puerto Rican was significantly different from how people thought of themselves in those communities, then it would show up in the interviews. We also did background research and asked questions that would allow us to draw some preliminary conclusions about social structures that might not be explicit in culturally held ideas. These efforts were productive, and we realized that Afro-Puerto Rican communities exist in socio-geographical "pockets" throughout the island—some in regions that have been historically tied to black people (coast on the eastern half of the islands) and others in tucked away neighborhoods in places that are not popularly known as "places where there are black people."

We also studied and relied on previous genetic ancestry research in Puerto Rico. The work by Martínez-Cruzado and colleagues (2001, 2005) is foundational and establishes island-wide patterns of genetic ancestry that have been confirmed by other researchers (Vilar et al. 2014). Another work by Via and colleagues, including Martínez-Cruzado (2011) was particularly relevant to our work for its attention to the geographical distribution of African ancestry throughout the island, which further explained Risch and colleagues' (2009) study showing how African ancestry has been a factor in assortative mating for Puerto Ricans living in Puerto Rico and New York. That is, these studies showed that African ancestry is important for understanding genetic stratification in Puerto Rico, and that historical processes of colonization and assortative mating explain that stratification. Nevertheless, the various works by Martínez-Cruzado and colleagues did not quite resonate with the way Afro-Puerto Ricans were describing their population. It is worth examining this lack of resonance because they reveal important differences in how genetic ancestry work may or may not serve as a tool for community empowerment. As anthropologists, we have professional responsibilities to both the communities and the colleagues with whom we work. However and as we saw in

Chapters 4 and 5, often there are particular aspects of research that are intricately tied to the sociocultural world in which research takes place. The definition of Puerto Rico as a population for genetic research is, in these previous works, tied to the United States census, which is based upon a nationalistic historical framework that has often been exclusive of Black Puerto Rican voices. Regardless, we assert that these studies served their respective purposes in highlighting a cross-section of genetic diversity as reflected by census populations. The use of census population, however, can be problematic specifically because of historic frameworks that silence or otherwise marginalized certain communities. Our criticism of this aspect of previous research is done with a spirit of respectful engagement and desire for future counter criticism of our work.

Barrios negros ("black barrios"), comunidades afro descendientes ("afro descendant communities"), gente negra ("black people"), los negros de *x* ("black people from *x*"). These are some of the terms that people with whom we talked around the island used to describe the many Afro-Puerto Rican communities that exist around the island. In some western parts of the island, a community is not strictly defined by geography since many Afro-Puerto Ricans who consider themselves part of a community live in different parts of towns. However, they have memories of past geographic spaces that were black, and they still gather in common spaces as Black people. In other parts of the island, like the barrio of San Antón in Ponce, the geographic concentration of Afro-Puerto Ricans is restricted to a few blocks. But even in that case, different families can tell you where people from San Antón have settled elsewhere in the region. Some of these places are hard to find since they exist at the margins of the formal urban infrastructure. But the people are there. Black people exist throughout the island, not just in Loíza. People in Loíza will readily point this out, and people outside of Loíza or a barrio like San Antón struggle to define their Blackness in geographical spaces that do not register in the national imagination as "traditionally Black" (Godreau 2015).

In all of these communities, nevertheless, people have a much easier time identifying as "negro" ("black"). In fact, in several interviews, research participants used almost identical phrasing in response to their racial identification, "pues negro, porque negro soy" ("well black, because I am black"). The first part of this phrase can and did have different intonations to either signal that being black is obvious or that choosing anything other than black is not an alternative. Because all research participants identify as Puerto Rican, their blackness was explained as much more than the African heritage that is part of

Puerto Rico's tri-cultural heritage. Being Taíno or being Spanish does not equate with the lived experiences of being Black in Puerto Rico, since the former are forms of heritage that they can own as Puerto Ricans but Blackness is not something that Puerto Ricans can own as non-Black people.

Another important cultural pattern that emerged in our interviews was intimate knowledge of local places and migrations between those places that span the whole island. For example, several families in Loíza described the origins of their ancestors in southern and western towns of the island. Others talked about their family coming from a black barrio or the black people of a certain town; and interestingly, many of these towns are in the mountain region—where, supposedly, black people have never really inhabited. In addition to the migration of people, there was also talk about the migration of musical traditions, instruments, and labor practices. The complexity of knowledge with regards to interisland migrations and social networks is also specific to the "traditionally" black town of Loíza. Historical differences between Piñones and Loíza are the most obvious. But even within Piñones and Loíza there are differences in how families have belonged in different spaces, migrated from different areas, and have different labor practices. Perhaps the most important form of cultural knowledge that is particular to local communities has to do with the attachment to land. From stories about the places where Black people gathered to intimate knowledge of environment, Afro-Puerto Ricans are clear not just about belonging to certain spaces, but having indigenous knowledge of where they have survived as Black people (cf. Giusti Cordero 1996, 2015). Enslavement is an important component of Afro-Puerto Rican pasts, but it is not necessarily defining of those pasts. Moreover, Maroon communities and communities of free blacks are also an important part of Afro-Puerto Rican heritage (cf. Moret 1984; Picó 1986).

The existence of small communities throughout the island, the sense of Blackness as a form of relatedness, and knowledge of migrations and landscapes are all factors that should immediately give us pause in defining Puerto Rico's population at the national level. For Afro-Puerto Rican voices tell us that there are socio-geographic spaces occupied by people who are related through family and racially embodied difference—these are the basic components of the working definitions of a population for ancestry research which we outlined above. However, in the work by Via et al. (2011), we are told African ancestry is high on the eastern part of the island and lowland regions, and that African ancestry is negligible on the western part of the island. Furthermore, they claim, the history of slavery and sugar plantations explains this pattern, and that one important historical marker—ports where slaves were imported—are not significant in explaining contemporary genetic ancestry patterns. Another argument explains that there was a lack of slaves or their descendants that moved into the interior mountainous regions, and the overall lack of movement around the island, " ... our results demonstrate that the descendants of slaves remained in the same areas where their ancestors resided 5–6 generations ago or moved to nearby locations" (2011, 6). But as we have shown above, these claims are directly contradicted by Afro-Puerto Ricans in our studies and previous historical research, which means that Via et al.'s interpretation of their genetic data is inadequate as a result of their limited recognition of Afro-Puerto Ricans people's lived experience.

The genetic ancestry science of these previous studies might go unquestioned if we do not pay close attention to how nationalisms represent forms of relatedness that do not concord with lived experiences. In the case of the studies cited above, we think good faith and good science characterized these studies. However, Via et al (2011) utilize the same data set sampled by Martínez-Cruzado et al. (2005). The latter adapted a random sampling strategy that was national in scope and utilized a combination of criteria, including population density, municipalities, and census tracts (2005, 133–34). Because Afro-Puerto Ricans in Loíza coincide with a municipality unit, their African ancestry is well documented and recognized in these studies. However, Afro-Puerto Rican communities in the west and the south—who have ancestry estimates similar to those in Loíza—do not register as populations, let alone living communities. The historical discrepancies between these studies and community histories is a little harder to explain since it is well established that places like Ponce (south) and Mayagüez (west) held high number of slaves in the eighteenth and nineteenth centuries (Díaz Soler 1970, 91; Centro de Investigaciones Históricas 1974, 180–82) and Afro-Puerto Ricans can often easily identify relatives in other parts of the island and have clear sense of when these relatives migrated. Afro-Puerto Ricans in Mayagüez, for example, are aware of the large presence of Black people in that part of the island as a result of both a major slave-importing port and slave labor in the region; moreover, most have stories of past family migrations to the south and eastern part of the island. These notions of relatedness, coupled with a lack of concentration in a geographical space, explain how researchers can miss Afro-Puerto Rican communities as populations for research.

Most importantly, genetic ancestry research in partnership with Afro-Puerto Ricans serves as a form of resistance to the exclusion of Afro-Puerto Ricans, who have been systematically marginalized in Puerto Rico. The resistance is rooted in the interests of communities who organize and advocate for a better future but also in the more robust scientific approach to genetic ancestry. Just as Anténor Firmin corrected a faulty anthropological model with a better anthropological model, genetic ancestry as resistance does not represent an alternative sampling methodology—it represents a more accurate and theoretically nuanced notion of population that takes race as a bioculturally embodied form of difference seriously. The nature of this seriousness is not prima facie political—it is serious as an empirically verifiable aspect of human experience. The political consequences of this science can better inform democratic deliberation on questions of equity and social justice (Torres Colón 2018).

References

Anderson, Benedict. 2006. *Imagined Communities: Reflections on the Origin and Spread of Nationalism*. Rev. ed. London: Verso.

Benn Torres, Jada, and Gabriel A. Torres Colón. 2015. "Racial Experience as an Alternative Operationalization of Race." *Human Biology* 87 (October): 306–12.

Brusi-Gil de Lamadrid, Rima, and Isar P. Godreau. 2007. "¿Somos Indígenas?" *Dialogo* Portada (marzo-abril): 10–11.

Centro de Investigaciones Historicas. 1974. *El Proceso abolicionista en Puerto Rico: Proceso y efectos de la abolicion: 1866–1896*. San Juan: Centro de Investigaciones Históricas, Facultad de Humanidades, Universidad de Puerto Rico, Instituto de Cultura Puertorriqueña.

Crawford, Michael H. 2007. *Anthropological Genetics: Theory, Methods and Applications*. Book, Whole. Cambridge; New York: Cambridge University Press.

Díaz Soler, Luis M. 1970. *Historia de la esclavitud negra en Puerto Rico*. San Juan: La Editorial, UPR.

Estevez, Jorge. 2008. "Amerindian MtDNA in Puerto Rico: When Does DNA Matter?" *Centro Journal* 20 (2): 219–28.

Fuller, Harcourt, and Jada Benn Torres. 2018. "Investigating the 'Taíno' Ancestry of the Jamaican Maroons: A New Genetic (DNA), Historical, and Multidisciplinary Analysis and Case Study of the Accompong Town Maroons." *Canadian Journal of Latin American and Caribbean Studies/ Revue Canadienne Des Études Latino-Américaines et Caraïbes* 43 (1): 47–78.

Gelman, Susan A., and Cristine H. Legare. 2011. "Concepts and Folk Theories." *Annual Review of Anthropology* 40 (1): 379–98. doi:10.1146/annurev-anthro-081309-145822.

Giusti Cordero, Juan A. 1996. "Labour, Ecology and History in a Puerto Rican Plantation Region: 'Classic' Rural Proletarians Revisited." *International Review of Social History* 41 (S4): 53–82. doi:10.1017/S0020859000114270.

———. 2015. "Trabajo y Vida En El Mangle: 'Madera Negra' y Carbón En Pinoñes (Loíza), Puerto Rico (1880–1950)." *Caribbean Studies* 43 (1): 3–71. doi:10.1353/crb.2015.0001.

Godreau, Isar P. 2015. *Scripts of Blackness: Race, Cultural Nationalism, and U.S. Colonialism in Puerto Rico.* Urbana: University of Illinois Press.

Haslip-Viera, Gabriel. 2008. "Amerindian MtDNA Does Not Matter: A Reply to Jorge Estevez and the Privileging of Taíno Identity in the Spanish-Speaking Caribbean." *Centro Journal* 20 (2): 228–37.

Inda, Jonathan Xavier, ed. 2005. *Anthropologies of Modernity: Foucault, Governmentality, and Life Politics.* Malden, MA: Blackwell Pub.

Krimsky and Sloan, eds. 2011. *Race and the Genetic Revolution: Science, Myth, and Culture.* New York: Columbia University Press.

Madrilejo, Nicole, Holden Lombard, and Jada Benn Torres. 2014. "Origins of Maroonage: Mitochondrial Lineages of Jamaica's Accompong Town Maroons." *American Journal of Human Biology* 27 (3): 432–437.

Martinez-Cruzado, J. C., G. Toro-Labrador, V. Ho-Fung, M. Estevez-Montero, A. Lobaina-Manzanet, D. A. Padovani-Claudio, H. Sanchez-Cruz, et al. 2001. "Mitochondrial DNA Analysis Reveals Substantial Native American Ancestry in Puerto Rico." *Human Biology* 73 (4): 491–511. doi:10.1353/hub.2001.0056.

Martínez-Cruzado, Juan C., Gladys Toro-Labrador, Jorge Viera-Vera, Michelle Y. Rivera-Vega, Jennifer Startek, Magda Latorre-Esteves, Alicia Román-Colón, et al. 2005. "Reconstructing the Population History of Puerto Rico by Means of MtDNA Phylogeographic Analysis." *American Journal of Physical Anthropology* 128 (1): 131–55. doi:10.1002/ajpa.20108.

Moret, Benjamin Nistal. 1984. *Esclavos profugos y cimarrones: Puerto Rico, 1770–1870.* San Juan: La Editorial, UPR.

Patterson, Orlando. 1970. "Slavery and Slave Revolts: A Socio-Historical Analysis of the First Maroon War Jamaica, 1655–1740." *Social and Economic Studies* 19 (3): 289–325. https://www.jstor.org/stable/27856434.

Picó, Fernando. 1986. "Esclavos, Cimarrones, Libertos Y Negros Libres En Rio Piedras, 1774–1873." *Anuario de Estudios Americanos; Sevilla* 43 (January): 25–33.

Risch, Neil, Shweta Choudhry, Marc Via, Analabha Basu, Ronnie Sebro, Celeste Eng, Kenneth Beckman, et al. 2009. "Ancestry-Related Assortative Mating in Latino Populations." *Genome Biology* 10 (11): R132. doi:10.1186/gb-2009-10-11-r132.

Sahlins, Marshall. 2000. *Culture in Practice: Selected Essays.* New York: Zone Books.

Stoneking, Mark. 2017. *An Introduction to Molecular Anthropology.* Hoboken, NJ: Wiley Blackwell.

Torres Colón, Gabriel A. 2018. "Racial Experience as Bioculturally Embodied Difference and Political Possibilities for Resisting Racism." *The Pluralist* 13 (1): 131–42.

Via, Marc, Christopher R. Gignoux, Lindsey A. Roth, Laura Fejerman, Joshua Galanter, Shweta Choudhry, Gladys Toro-Labrador, et al. 2011. "History Shaped the Geographic Distribution of Genomic Admixture on the Island of Puerto Rico." *PLoS One* 6 (1): e16513. doi:10.1371/journal. pone.0016513.

Vilar, Miguel G., Carlalynne Melendez, Akiva B. Sanders, Akshay Walia, Jill B. Gaieski, Amanda C. Owings, and Theodore G. Schurr. 2014. "Genetic Diversity in Puerto Rico and Its Implications for the Peopling of the Island and the West Indies." *American Journal of Physical Anthropology* 155 (3): 352–68. doi:10.1002/ajpa.22569.

Conclusion
Our stories, our past

In this book we have argued that genetic ancestry has no inherent social meaning, and that we must pay careful attention to how people think about the past and relatedness before we can understand how people experience genetic ancestry technology and information. We have also argued for a biocultural approach to the study of embodied difference, particularly race that leads to a more nuanced understanding of genetic ancestry as not only problematically interrelated to folk theories about race but also a form of racial experience. In turn, how we think about race (both as a sociocultural phenomenon and a subject of social scientific inquiry) is grounded in a broader historical and contemporary context of anthropological practice. We have argued that this context of anthropological inquiry, if rooted in the legacy of Anténor Firmin, should be inclusive of a science and social science that is fundamentally anti-racist. Such an anthropological approach can lead to the use of genetic technology as a tool of empowerment and anti-racism. The arguments put forth in this book also lead us to believe that an integrative anthropology is a well-suited home for scientific inquiry into biocultural dimensions of race and anti-racist anthropology. Cultural anthropologists, biological anthropologists, and broader audiences can only benefit from more nuanced perspectives that move beyond declarations like "race is not biologically real" or "race is a social construct." Indeed, our research into genetic ancestry demonstrates that there is much room for growth in how anthropologists and other scholars continue to engage racial experience as research and as a prism for anti-racist struggles.

One of the similarities between genetic ancestry consumers, Anténor Firmin, contemporary anthropologists, critics of genetic ancestry products, and various social actors involved in anti-racist struggles is that they have been interested in the present and future relevance of how we think about our ancestors. Anthropologists, of course, have a disciplined approach to how we study humans' past and relatedness; but they do not have particular stake or higher intellectual ground on the significance of our ancestry. Is the value of anthropology in highlighting our similarities, differences, or both? If both, then how do we find significance in knowing—to use our own ancestries as examples—that all humans share common ancestors in Africa, but that our admixture estimates result from a history of colonization and genocide in the Caribbean? Ideally, knowledge that all humans have common ancestors in Africa should be a good reason for preventing historical processes, such as colonial processes in the Caribbean, from ever taking place. However, people throughout the Caribbean and other parts of the world continue to live in marginalized conditions as a direct result of those colonial processes. Therefore, knowledge about our deep human ancestry becomes much less pressing and relevant than African admixture estimates for Afro-Puerto Ricans, Accompong Maroons, or Santa Rosa First Peoples Community.

Genetic ancestry is increasingly part of our stories and our past—sometimes more important, sometimes not important at all. But since the significance of our stories and our past are differently relevant to how we move about the world, fight for justice, and relate to one another, it is crucial that we listen to each other's stories, others' pasts; and that we are tolerant in our misunderstandings, demanding in our quests to be heard, and forgiving in finding resolutions to past injuries. Or, as Firmin concludes,

> Throughout all the struggles that have afflicted, and still afflict, the existence of the entire species, one mysterious fact signals itself to our attention. It is the fact that an invisible chain links all the members of humanity in a common circle. It seems that in order to prosper and grow human beings must take an interest in one another's progress and happiness and cultivate those altruistic sentiments which are the greatest achievement of the human heart and mind.
>
> (Firmin 2000: 450)

This from a Haitian writing in Paris in 1885.

References

Firmin, Joseph-Anténor. 2000 [1885]. *The Equality of the Human Races: (Positivist Anthropology)*. New York: Garland Pub.

Index

Printed in the United States
by Baker & Taylor Publisher Services